MOBILE CLOUDS

MOBILE CLOUDS

EXPLOITING DISTRIBUTED RESOURCES IN WIRELESS, MOBILE AND SOCIAL NETWORKS

Frank H.P. Fitzek
Aalborg University, Denmark

Marcos D. Katz
University of Oulu, Finland

This edition first published 2014

Registered office
John Wiley & Sons Ltd, The Atrium, Southern Gate, Chichester, West Sussex, PO19 8SQ, United Kingdom

For details of our global editorial offices, for customer services and for information about how to apply for permission to reuse the copyright material in this book please see our website at www.wiley.com.

Library of Congress Cataloging-in-Publication Data

Fitzek, Frank H.P.
Mobile clouds: exploiting distributed resources in wireless, mobile and social networks / Frank H.P. Fitzek, Marcos D. Katz.
pages cm
Includes bibliographical references and index.
ISBN 978-0-470-97389-9 (hardback)
1. Cloud computing. 2. Mobile computing. I. Katz, Marcos D. II. Title.
QA76.585.F58 2014
004.67′82–dc23

2013030750

A catalogue record for this book is available from the British Library.

ISBN: 978-0-470-97389-9

Set in 10/13pt Times by Aptara Inc., New Delhi, India

1 2014

To Lilith and Samuel!

Contents

Part Four GREEN ASPECTS OF MOBILE CLOUDS

Part Five APPLICATION OF MOBILE CLOUDS

Part Six MOBILE CLOUDS: PROSPECTS AND CONCLUSIONS

Foreword

The penetration of mobile technology into our society in recent years is remarkable. It has enabled enormous levels of always–available connectivity to the world's population with untold benefits and capabilities. This book on Mobile Clouds lays it all out. It is written at an easily accessible level for engineers, researchers and students without the burden of heavy mathematics, but rather with a superb descriptive approach that encourages full understanding of the key issues, the basic solutions, the advantages of those solutions, the direction in which the field is moving, and a presentation of its impact. This is a highly readable, intuitively pleasing and most enjoyable presentation of the emerging world of Mobile Clouds.

The trajectory of the development of Mobile Clouds has been one of wireless communications leading to mobile telephony, which has evolved into always–available voice and data access. The tremendous success of these technologies is placing severe strains on the underlying resources needed to continue the growth and deployment of new users, new applications, and new services. In response, as this book explains, there is a growing need for sharing of resources while at the same time improving the efficiency of spectrum use and energy consumption. The seeds for these improvements came from two parallel developments in the early history of mobile access, both beginning in the early 1970's. One development is that with which the public is so familiar, namely, the rise of mobile voice access which led to the centralized point–to–point architecture of the network operator directly communicating with the mobile device. This led to the recognition that improved spectrum and energy efficiency would result from the introduction of smaller and smaller cell sites. The other development, far less familiar to the public, is the study of the distributed architecture of multi–hop mesh network communications in which each node became a part of what we now recognize as the Mobile Cloud. There is great promise now in the growth of a hybrid distributed/centralized architecture, which exploits the best of both architectures. In this architecture, the technology of network coding plays a valuable role and, true to their form, the authors provide a lucid and intuitive description in a full Chapter devoted to this important topic.

Rounding out this work, the application of mobile clouds focuses on various forms of cooperation in social networks, such networks serving as an important driver of Mobile Cloud growth. As we move into the future, we are then introduced to the growth of traffic due to Machine–to–Machine communication as well as the huge move to the Internet of Things. We have moved into a new era in which the embedded devices of the Internet of Things, in addition to the intelligent software agents that populate the Internet, are generating more Internet traffic than are humans. As we race into this future, the need for an understanding and appreciation of the emergence and role of Mobile Clouds is critical. This book provides what you need to know.

Leonard Kleinrock
Distinguished Professor, UCLA, Computer Science Department
3732G Boelter Hall, Los Angeles, California 90095

Preface

Putting Mobile Clouds into Context

Wireless and mobile communications have rapidly evolved, offering today high–speed connectivity and advanced services to a continuously rising number of subscribers on the move. Currently, there are well above seven billion mobile subscribers worldwide and ITU predicts that penetration figures will exceed 100% in 2014. Since the turn of this century the developments in wireless and mobile communication systems became faster, particularly in access networks, mobile devices and service technologies. The principal design goals of mobile networks have been to increase data throughput and energy efficiency. These goals have well been achieved by cellular networks through several technology generations.

Current cellular networks can establish data connections at rates that were unthinkable a decade ago, and in many cases the speeds are comparable to what is today offered by wired networks. Two present trends in communications are creating new demands and challenges to current mobile and wireless communications technology. These are the current rapid development of social networking as well as the emergence of Machine–to–Machine (M2M) and Internet of Things (IoT) technologies.

The patterns of how people communicate and socialize have changed and continue evolving, mostly inspired and supported by the Internet. Ubiquitous connectivity is now a reality, people can be connected to each other, access information and distribute their own content regardless of their location. The emergence of technology–based social networks has further changed the way people live and interact. The Internet is the enabling platform for social networking at any scale, local or global. Today social networking increasingly takes place from mobile devices and consequently the role of wireless and mobile communication networks becomes even stronger. In the future the interplay of social and mobile networks will boost the ideas of shareconomy.

Social interaction involves not only the creation of individual (person–to–person) links but also establishing one–to–many and many–to–many connections. In addition to user–controlled mobile devices, machines and ultimately things will become nodes of communication networks, promptly increasing the number of nodes to be potentially interconnected by several orders of magnitudes. It has been predicted that in the third decade of this century there could be as many as several trillion communication–enabled nodes on this planet. Current communications networks cannot scale efficiently to support the large networks of the future.

Spectral– and energy–efficiency of current network solutions have long been identified as significant roadblocks in the development path. It is a well–known fact that spectrum bands allocated to mobile communications are very much limited and expensive. Provision of high data rates to support wireless delivery of rich content swiftly increases bandwidth requirements. Moreover, when these requirements are mapped into the projected growth of the node base, the results speak by themselves: spectral efficiency of future networks need to be greatly boosted at both link and network level.

Energy efficiency is another extremely important challenge of future communication often referred to as *green* communication [1]. On the infrastructure side the amount of energy needed to provide access services to mobile users is significantly high. A single network operator easily spends several million Euros per year in electricity costs to provide its network access services to a middle–size city. When these figures are scaled up to country or global basis, the economical and environmental impacts are certainly notable. In the other end of the communications value chain, energy efficiency of mobile devices is also an important factor daily experienced by users and highly significant to mobile device manufacturers. Long operating times of portable devices is a highly desired capability for discerning users, and a key competitive feature offered by manufacturers. Nowadays access to wireless communication systems is not limited by coverage any longer, but by the operational time of the mobile device. A trend that we had already predicted in [2].

In summary, one of the key challenges resulting from the increasingly richer social interaction between people and the advent of machine communications is the explosive increase in the use of resources of the communications networks. Another challenge faced by communications networks is the provision of low–latency end–to–end services. Real–time services such a video calls set stringent requirements on the involved communications delay. Applications based on machine communications will further set the requirements bar higher, calling for even shorter delays [3]. Current delay figures, in the range of hundred milliseconds, are expected to be reduced by one or two orders of magnitude. Solutions to cope with the aforementioned challenges can be developed at different levels. A straightforward approach would involve developing sophisticated air interfaces, the somewhat trivial but highly challenging approach that has been exploited along the development of the mobile technology generations. Structural changes at network level can have a deeper impact on the way information flows in the network, and hence determining how network resources are used and having an effect on the involved latency.

Mobile networks architecture has largely remained unchanged since its introduction. Even though this centralized access approach has proved to work properly and is the basis of today's mobile networks, it is clear that it is does not use efficiently the available radio resources. In recent years extension of the cellular architecture have been put into use, including the emergence of cooperative approaches such as relaying (multi–hop) techniques. Furthermore, recently the concept of Device–to–Device (D2D) has taken off, and it is currently a widely studied approach in the LTE–A (Long Term Evolution–Advanced) standardization process. On the other hand, wireless networks have made use of less rigid access topologies, supporting by design the establishment of direct peer–to–peer links as well as centralized connections to access points. Mobile clouds, introduced and studied in this book, build a bridge between mobile and wireless communication networks, by creating a composite centralized–distributed access architecture. One of the purposes of the mobile clouds' hybrid topology is to exploit the best of both worlds, the wide access and simple centralized manageability on one hand, and the flexible, rapid access of local networks on the other hand. One of the major trends is latest years is the emergence of cloud–based services.

Cloud solutions are implemented either on geographically distributed cloud nodes or they can be based on lumped approaches, concentrated for instance on a single powerful node. In any case users, fixed or mobile, can access the cloud regardless of their physical location. This model works well but, when considering mobile users, the practical solutions, e.g., access networks, use considerable amounts of radio resources. The problem of inefficient usage of resources becomes more pronounced the more mobile nodes are involved, like in cases of social networking. Platforms providing cloud services are deep inside the backbone network and far away from access networks. In addition to excessive consumption of energy and spectrum, accessing remote clouds inevitably means high associated delays. The closer the cloud is to the mobile user, the more efficiently the services can be wirelessly accessed. In addition to the mentioned conventional clouds there is a need to have cloud–based operations closer to the user. Such trends are already visible as in the developments supporting Device–to–Device interaction, a key building capability of 5G networks. This book is devoted to introducing and discussing the concept of mobile clouds.

Mobile Clouds

As we will define later, a mobile cloud is a cooperative arrangement of dynamically connected nodes sharing opportunistically resources. Both mobile and wireless network technologies are opportunistically combined to achieve a number of possible goals. Mobile clouds can be considered as an evolutive step towards bringing cloud–based services closer to the user themselves. In fact, users can become central players as their devices become nodes of a mobile cloud. Mobile clouds offer unique and attractive gains in three main domains: namely performance, resource efficiency and resource exploitation. Mobile clouds have the potential to enhance key link and network performance measures, including supported data throughput, latency, reliability, security as well as capacity and coverage. Mobile clouds can also provide practical solutions with high spectral and energy efficiency. In particular, the impact of mobile clouds on energy consumption of mobile devices, base stations or access points is highly significant and mobile clouds can be seen as one of the enabling technologies for future green networks. One of the most exciting applications of mobile clouds is as a platform for sharing the distributed resources residing in the cloud. A large number of resources (physical or intangible) can be shared in many manners using a mobile cloud as a flexible and efficient exchanging platform.

This book advocates for mobile clouds as the upcoming mobile communication platform of the future, extending the commonly known point–to–point connection between network operator and mobile device. Parts of this development have been introduced already in [2, 4] by our world–class colleagues but here we present the state–of–the–art with recent developments and future developments on the horizon.

Aims of the Book

The main aim of this book is motivating readers on the potential of mobile clouds for implementing a large number of possible solutions needed or emerging in our present and future mobile and wireless world. Given that mobile clouds as such is a relatively new concept, a complete account of mobile cloud technology is not yet available. The goal of this book is to serve as an inspiring source for researchers, developing engineers and students interested in solutions for future wireless and mobile networking. The book describes mobile clouds and their uses from the above–mentioned goals. Many inspiring examples are presented and discussed. In some cases precise analytical models are presented and explained, accompanied with numerical results showing concrete figures of the achievable gains. The authors also include some practical information on mobile clouds test–beds, showing the practical applicability of this concept.

Organization of the Book

This book is organized in six main parts with eleven chapters. For newcomers to the field of mobile clouds we propose they read books chapters in sequential order. The experienced reader can directly go to the chapters that are of greater importance to the reader. Each chapter is self contained, which results in some planned overlap.

- Part I includes three chapters. The motivation chapter is describing the wireless and mobile context, while the second chapter is introducing the mobile cloud concept giving several definitions. The third chapter is identifying sharable resources on a mobile device listing several examples.
- Part II deals with enabling technologies for mobile clouds. Chapter 4 lists current wireless technologies and their capability to build mobile clouds. The fifth chapter is introducing network coding, which is a key technology for mobile clouds allowing flexible design with low resource usage. The sixth chapter describes mobile cloud formation and maintenance.
- Part III contains two chapters explaining cooperative principles in nature and the social mobile cloud concept. In this book we envision mobile clouds to be built up by individuals who need to be convinced that cooperation in a mobile cloud is beneficial for all participants.
- Part IV focuses on green aspects of mobile clouds showing potential energy saving gains from the theoretical point of view for different application scenarios.
- Applications of mobile clouds are presented and discussed in Part V. Here the ongoing activities are described mainly from the mobile app perspective.
- Finally Part VI discusses prospects of mobile clouds and draws conclusions.

Frank H.P. Fitzek
Aalborg, Denmark

Marcos D. Katz
Oulu, Finland

References

[1] H. Zhang, A. Gladisch, M. Pickavet, Z. Tao, and W. Mohr. Energy efficiency in communications. *IEEE Communications Magazine*, 48(11):48–49, 2010.

[2] F.H.P. Fitzek and M. Katz, editors. *Cooperation in Wireless Networks: Principles and Applications–Real Egoistic Behavior is to Cooperate!* ISBN 1-4020-4710-X. Springer, April 2006.

[3] G. Fettweis. A 5G Wireless Communications Vision. *Microwave Journal*, December 2012.

[4] F.H.P. Fitzek and M. Katz, editors. *Cognitive Wireless Networks: Concepts, Methodologies and Visions Inspiring the Age of Enlightenment of Wireless Communications.* ISBN 978-1-4020-5978-0. Springer, July 2007.

Acknowledgements

We, Frank and Marcos, would like to thank everybody that has inspired us throughout the process in making this book.

The idea of writing a book on mobile clouds was conceived in a cloudless day on a Honolulu beach, during Globecom 2009. We discussed with Wiley's Mark Hammond about our idea and promised him to start immediately working on the project. Several Globecom conferences and countless clouds passed by, and now our mobile cloud book is being introduced at Globecom 2013. We would like to thank Mark for his immense patience and professional support through the writing period. We would equally like to thank Wiley's Liz Wingett and Anna Smart for very similar reasons.

Frank would like to thank his team and colleagues at Aalborg University and colleagues around the globe for the support over the last decade. Aalborg University has provided a fertile ground for my research and I always found motivated colleagues to collaboratively research on mobile clouds. Special thanks to Muriel Médard for the fruitful discussion on network coding and support over the last years. I would like to thank Hassan Charaf for his long lasting cooperation and for the successful exchange of students over the last years. Thank you to Daniel Lucani for his help in proof reading and valuable comments. Thanks to Peter Vingelmann for his work on multimedia sharing on Apple products. Thanks to Kirsten Nielsen for organizing our work and life. Special thanks to Morten V. Pedersen for his long lasting cooperation and friendship over the last years. He is the mastermind of our code examples and I would like to thank him for his unbreakable will to *change the code base* for a better future. Also, our financial support over the years shall not be forgotten. Parts of this book were partially financed by the CONE project (Grant No. 09-066549/FTP) granted by Danish Ministry of Science, Technology and Innovation. Further funding was received by the Green Mobile Cloud project granted by the Danish Council for Independent Research (Grant No. 10-081621). Also, thanks to our supporters from the ENOC project in collaboration with Renesas and Nokia, Oulu.

Marcos would like to thank Centre for Wireless Communications and University of Oulu, Finland for providing me invaluable support as well as an inspiring working atmosphere. Marcos would also like to thank his colleagues and students for their support and enthusiasm. A particular warm thanks to my closest research team Timo Bräysy, Zaheer Khan, Hamidreza Bagheri, Bidushi Barua, Muhammad Ikram Ashraf, Helal Chowdhury and Syed Tamoor-ul-Hassan. Professor Babak Hossein Khalaj and Mohammad Javad Salehi from Sharif

University of Technology, Iran, are kindly acknowledged for their cooperation in this subject. Professor Miguel A Cabrera and Fernando Miranda Bonomi from National University of Tucumán, Argentina are also acknowledged for their efforts during our ongoing cooperation. The inspiring discussions with Kari Horneman (Nokia Solutions and Networks), Pavel Loskot (Swansea University) and Pekka Sangi (University of Oulu) are greatly appreciated. A well–deserved thanks also to Hanna Saarela, Kirsi Ojutkangas and Eija Pajunen, our always–smiling administrative staff at Centre for Wireless Communications, for their charming support and help. Tekes, the Finnish Funding Agency for Technology and Innovation is acknowledged for its generous financial support through the SANTA CLOUDS, COIN and INDICO research projects. The European Celtic–Plus initiative, together with Tekes are acknowledged for their support on the Green–T project. Marcos is also grateful to numerous colleagues across the world with whom he has had the honor to work with in many areas of wireless and mobile communications.

Abbreviations

3GPP	Third Generation Partnership Project
API	Application Programming Interface
ARQ	Automatic Repeat reQuest
AWS	Amazon Web Services
BATMAN	Better Approach To Mobile Ad Hoc Networking
BB	Base Band
BOM	Bill Of Materials
CA	Collision Avoidance
CATWOMAN	Coding Applied To Wireless On Mobile Ad Hoc Networks
CCS	Cooperative Control Server
CDMA	Code Division Multiple Access
CPU	Central Processing Unit
CS	Circuit Switched
CSD	Circuit Switched Data
CSMA	Carrier Sense Multiple Access
CTS	Clear To Send
CUHD	Cellular Uplink Hybrid Downlink
D2D	Device–to–Device
DCF	Distributed Coordination Function
DRM	Digital Right Management
DS	Direct Sequence
DSP	Digitial Signal Processing
DTN	Delay Tolerant Networks
DVB	Digital Video Broadcasting
DVD	Digital Video Disc
DVS	Dynamic Voltage Scaling
EDGE	Enhanced Data for GSM Evolution
EDR	Enhanced Data Rate
EGPRS	Enhanced GPRS
FEC	Forward Error Correction
FH	Frequency Hopper
FTP	File Transfer Protocol

GB	Gigabyte
GMSK	Gaussian Minimum Shift Keying
GPRS	General Packet Radio Service
GPS	Global Positioning System
GPU	General Processing Unit
GSM	Global System for Mobile communication
HSCSD	High Speed Circuit Switched Data
HSDPA	High Speed Downlink Packet Access
HSPA	High Speed Packet Access (3G)
HSPA+	High Speed Packet Access (evolved 4G form)
HTML	HyperText Mark-up Language
HUCD	Hybrid Uplink Cellular Downlink
HUD	Hybrid Uplink and Downlink
ICT	Information and Communication Technology
IEEE	Institute of Electrical and Electronics Engineers
IoT	Internet of Things
IP	Internet Protocol
IPR	Intellectual Property Rights
IPTV	Internet Protocol Television
IR	Infra Red
IrDA	Infrared Data Association
ISM	Industrial Scientific Medical radio frequency band
ISO	Institutional Organisation for Standardization
ITU	International Telecommunication Union
LAN	Local Area Network
LED	Light Emitting Diode
LIPA	Local IP Access
LTE	Long Term Evolution
LTE–A	Long Term Evolution Advanced
M2M	Machine–to–Machine
MAC	Medium Access Control
MANET	Mobile Ad Hoc Network
MB	Megabyte
MIMO	Multiple Input Multiple Output
MTU	Maximum Transmission Unit
NFC	Near Field Communication
OFDM	Orthogonal Frequency Division Multiplex
OMC	Overlay Mobile Cloud
OSI	Open Systems Interconnection
PC	Personal Computer
PDA	Personal Digital Assistant
PSK	Phase Shift Keying
QoE	Quality of Experience

QoS	Quality of Service
QR	Quick Response
RF	Radio Frontend
RFID	Radio Frequency Identification
RLNC	Random Linear Network Coding
RTS	Ready To Send
SDK	Software Developer Kit
SIM	Subscriber Identity Module
SMS	Short Message Service
SNR	Signal–to–Noise Ratio
SRMC	Short–range Mobile Cloud
SSID	Service Set Identifier
TDMA	Time Division Multiple Access
TMC	Traffic Message Channel
UMTS	Universal Mobile Telecommunications System
USB	Universal Serial Bus
UWB	Ultra–Wide Band
VLC	Visible Light Communication
VoIP	Voice over Internet Protocol
WBAN	Wireless Body Area Network
WiMAX	Worldwide Interoperability for Microwave Access
WPAN	Wireless Personal Area Network
WLAN	Wireless Local Area Network
WSN	Wireless Sensor Network
WWI	World War I
WWII	World War II
WWRF	Wireless World Research Forum
XOR	The Inequality function

Part One

Mobile Clouds: Introduction and Background

1

Motivation

Inventions have long since reached their limit, and I see no hope for further development.
Julius Sextus Frontinus, highly regarded Roman engineer, 1st century A.D.

This chapter serves as a motivating introduction to the subject of this book: mobile clouds. A brief account of the evolution of mobile and wireless communications is presented from the point of view of mobile devices as well as communication networks. Mobile clouds can be considered as the result of the evolution and merging of mobile and wireless communications technologies. These initial pages will shed some light on some historical developments leading to the concept of mobile clouds.

1.1 Introduction

Untethered communications, omnipresent and fundamental in today's hyper-connected world, evolved rapidly in the last decades. The impact on our lives is so deep that it is hard to imagine how difficult it would be living now without the informational and social connectivity, freedom as well as flexibility brought by wireless communications technology. In this introduction we briefly discuss the evolutionary development of wireless communications until the present, from networks and mobile devices points of view. This overview will provide some useful and motivating background information before focusing on mobile clouds. Two evolutionary paths characterize untethered communications, the developments in *wide–area communications* on one hand, and the developments in *short–range communications* on the other hand. The former can be denominated the *mobile path*, while the latter is the *wireless path*, due to the fact that typically *mobile communications*, and *wireless communications* are the terms used for wide–area and short–range technologies, respectively. Radio broadcasting, the very first example of wide–area communications, started to be developed at the turn of the 20th century. WWI and WWII provided an immense thrust to the development of radar and communications

Mobile Clouds: Exploiting Distributed Resources in Wireless, Mobile and Social Networks, First Edition.
Frank H.P. Fitzek and Marcos D. Katz.

technology. The further developments in solid–state components resulted in miniaturization, made possible implementation of complex systems and gave birth to the era of truly portable communications equipment. The first urban mobile communications systems were deployed as early as in the latest 1940's. Single powerful base stations with high–rise antennas were initially used to provide access to areas with radius of up to some 50km. Already at that time scarcity in the available spectrum was identified as an issue and Bell Labs proposed the idea of covering large geographical regions by using a number of smaller service areas. Further developments in the upcoming decades led to the introduction of basic cellular systems for public and private use in the 1970's. Most of this pioneering work took place in the US but in the next decades Europe and Japan developed also their own commercial cellular systems. The cellular concept, based on frequency reuse in smaller coverage areas, or cells, allowed city–wide support of a large number of users. Through the 1980's until the present day four generations of cellular systems were developed, such that 2G, 3G and the rather recently introduced 4G coexist today. Requirements for higher supported data rates and network capacity led to a gradual reduction of cell sizes, typically up to few tens of kilometers in *macro–cells*, few hundred meters to few kilometers in *micro–cells* and from meters to a few hundred meters in the case of *pico–cells*. Certainly cell size is also related to mobility, large cells support higher degrees of mobility with the need for frequent handovers to adjacent cells. Providing untethered connectivity over short distances has also proved to be highly important, if not absolutely necessary, to a great deal of applications and in many practical scenarios. Over the last two decades a large number of communication technologies for short–range communications were developed fulfilling the demands for local wireless connectivity to computers, home and office appliances and other portable, movable or fixed equipment. This parallel development, the aforementioned wireless path, produced a very eclectic range of communications technologies covering from millimeters to a few hundred meters. Examples of short–range communications include wireless local area networks, (WLAN), wireless personal area network (WPAN), wireless body area network (WBAN), wireless sensor networks (WSN), radio frequency identification (RFID) and near field communications (NFC). Besides radio communication there is also optical communication, especially visible light communication (VLC). As compared to the developments in wide–area communications, focused mostly on overlay cellular networks operating on a centralized manner, short–range communications is a highly fragmented development arena, technology–, applications– and architecture–wise. The industry behind wide–area cellular and short–range communication fields are typically different. Large telecom manufactures back the former, whereas a diverse array of technology industry, with computer industry having the largest share, being behind the eclectic solutions existing for short–range communications. As we are moving towards a highly integrated mobile and communications era, the division between industry supporting cellular and short–range communications becomes blurred. Stretching from millimeters ranges of to hundreds of kilometers, wireless communications today consists of a large collection of different technologies omnipresent in our life. Figure 1.1 illustrates current representative mobile and communications approaches as a function of their typical ranges. Broadly speaking short–range and wide–area cellular communications remain today the main two approaches to untethered communications.

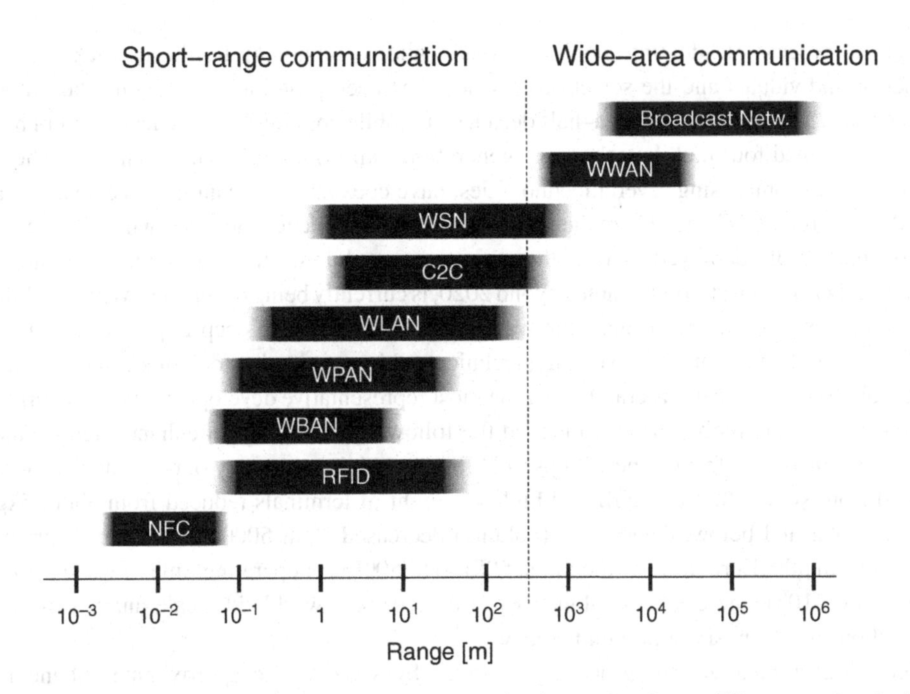

Figure 1.1 The realm of wireless and mobile communications today: from millimeters to hundreds of kilometers.

1.2 From Brick Phones to Smart Phones

Personal computers, Internet and mobile communications are among the most rapidly adopted technologies in history. In particular, the emergence and further popularization of mobile communication technologies are truly remarkable and unique achievements. Today, after a quarter of a century since the inception of mobile communications, the worldwide penetration of mobile and wireless communication devices exceeds 86% as given in [1]. Connectivity is seen today as an indispensable commodity, or even more, as a basic right of each individual. Mobile devices provide wireless access, making possible portable connectivity in most of the scenarios where people live, work and spend their free time. The outstanding development of mobile communications can be seen as the result of huge global research and development efforts by related industry, academia and regulators. Envisaging this rapid development in this area has always been a real challenge. Even the most optimistic forecasts were short to predict the colossal growth of mobile communications. In 1997 it was estimated that by 2010 there would be from one to two billion mobile subscribers [2, 3], whereas in 2006 such figure was estimated to be three billion [4]. The actual figure in 2010 well exceeded the five billions. In a few years from now (2014) the worldwide penetration is expected to reach or even exceed 100%.

These impressive figures are just one part of the story. Mobile and wireless communications have changed radically the way people communicate with each other and access information. And more changes will certainly follow. The impact of mobile communications on how people socialize, work, retrieve information, do business and entertain themselves is really enormous.

The global process of adopting mobile communications technology has been quick and its impact on individuals and the society as a whole has been profound, far beyond the initial expectations. The so far two–and–a–half decades of mobile communications development has basically spanned four mobile technology generations, known as 1G, 2G, 3G and 4G. These generations, encompassing several technologies, have coexisted and continue to coexist on a global scale. Today, 2G and 3G are the most widely used mobile technologies while 4G, being at this time rapidly deployed, will be the mainstream mobile technology in the near future. Moreover, 5G, aiming at a time-frame beyond 2020, is currently being developed. While mobile communications continues to shape the way that people live, such deep impact would have not been possible without the outstanding technical achievements that took place in the rather short mobile communications era. Among the most representative developments that occurred in the past 25 years mobile users witnessed the following technological enhancements: data rate support increased from some 100bps to 1Mbps and higher; memory onboard devices was boosted from some 1MB to 32GB and higher; weight of terminals reduced from about 5kg down to 100g and below; device size (volume) decreased from 5000cm^3 to 50cm^3; prices dropped from 5000Euro down to the range 50Euro to 500Euro; operating time saw a ten–fold decrease (1h–10h) whereas the total number of devices on a worldwide scale jumped from a few millions to nearly six billion units today.

Figure 1.2 summarizes these accomplishments by showing the approximate enhancing factors of key capabilities of mobile communications devices. Another major development in the evolution of mobile devices is the pronounced change in the nature of the mobile devices themselves. A large part of the mobile phone era has been characterized by devices designed just to provide basic connectivity (voice and data), with little or no available resources onboard

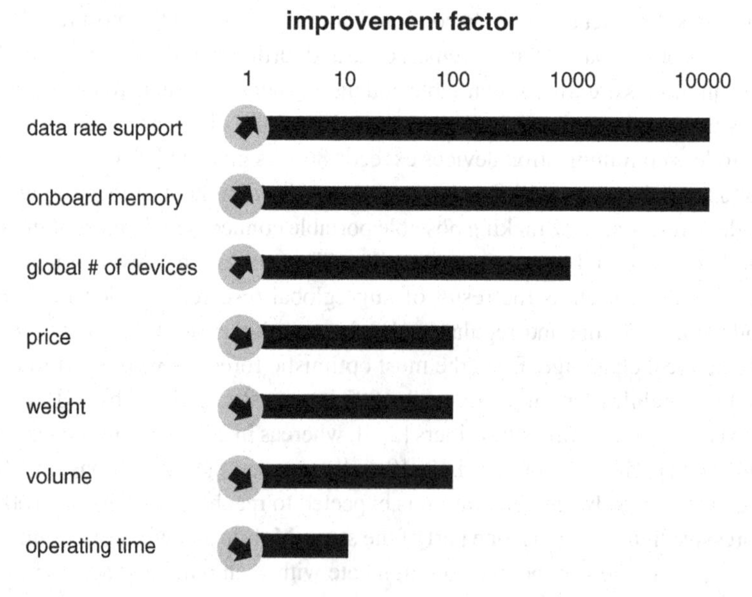

Figure 1.2 From brick phones to smart phones: mobile device evolution in the last 25 years (1985–2010).

for uses other than communications. Today, with the advent of smart phones, mobile users can enjoy sophisticated multipurpose devices, that can be seen as part of a large wireless ecosystem. Current devices have a great amount of resources on board, such as powerful processors, large mass memory, an increasingly number of sensors, a multiplicity of complementing air interfaces, advanced imaging components such as high resolution image sensors and displays. Another relatively new but immensely powerful extension to mobile phones is the development of mobile applications, *apps*, inexpensive pieces of software that can be easily downloaded bringing new capabilities to the devices. In 2011 more than 30 billion apps were downloaded into mobile phones. Mobile phones evolved from being closed–systems to become the flexible open–platforms of today. Indeed, the first generations of mobile phones were largely unchangeable, fixedly–programmed at factory, and with minimum or nonexistent support for updates or extensions. Today, the term *mobile phone* to a great extent does not accurately reflect the state of technology, *mobile device* being a more representative denomination for the current highly flexible, programmable and customizable wireless multifunction devices, that also work as mobile phones. Certainly, equally striking though less perceptible to users is the evolution undergone by the mobile communications networks, essential to match the high performance capabilities of wireless devices. Cellular networks development has been and continues to be focused on enhancing key performance figures such as supported data throughput, network capacity, quality of service, latency, reliability and coverage.

1.3 Mobile Connectivity Evolution: From Single to Multiple Air Interface Devices

In this section we shed some light on the development of the air interfaces used for mobile communications. As voice communications was the only capability of early mobile communication systems, relatively simple air interfaces were used, first based on analog designs (e.g., 1G) followed by digital approaches (e.g., 2G and beyond). The introduction of digital communications allowed naturally transfer of data, and this was first capitalized with short message services. Typically, mobile devices have had a single air interface providing just connectivity through cellular access. This simple initial approach is still widely in use today, particularly in the low–cost device segment. Such relatively simple air interface did provide low throughput connectivity, supporting initially voice and very low rate data transfer. Figure 1.3 (left side) depicts a representation of such a low–rate one–dimensional (i.e., single air interface) connectivity approach. As requirements for higher data rate support, larger coverage, and improved reliability increased, advanced air interfaces as well as networks were developed, exploiting a number of advanced techniques. These sophisticated air interfaces employed for instance different spatio–temporal processing techniques, such as diversity, beamforming and spatial multiplexing techniques. Multi–antenna approaches, known collectively as MIMO techniques (Multiple Input Multiple Output), are effective to increase data throughput, coverage and capacity, though, as a whole, the performance–complexity trade–off of MIMO techniques does not always lead to attractive engineering solutions. Moreover, advanced network architectures based on cooperative principles, e.g., multi–hop techniques, were introduced to

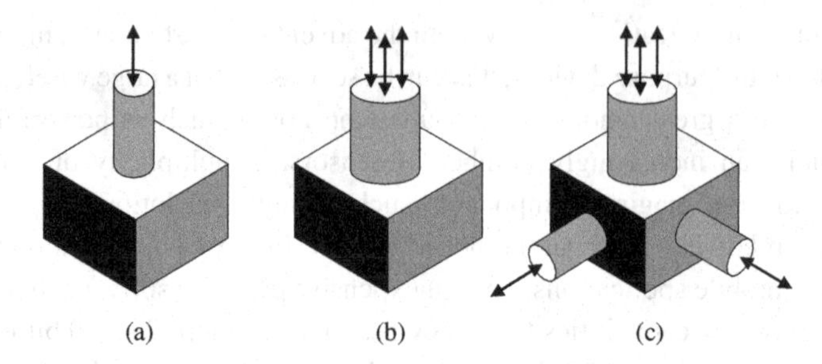

Figure 1.3 From brick phones to smart phones to the basic brick of mobile clouds: Air interface dimensionality, (a) single air interface; (b) single high–performance air interface and (c) multiple multi-dimensional air interfaces.

enhance performance and extend coverage. The generalized exploitation of radio resources (e.g., time, space, frequency) resulted in remarkable enhancements in performance at link and network levels. Figure 1.3 (middle) shows also a one–dimensional air interface approach as in Figure 1.3 (left), but supporting higher data throughputs due to the use of the aforementioned techniques.

Since the beginning of the mobile communications era, mobile devices have made use of rather simple centralized access architecture, connecting them to one or more base stations, directly or through repeaters. An interesting fact is that early wireless devices were equipped with additional wireless connectivity ports, notably optical air interfaces [e.g., Infrared Data Association (IrDA)] for very–short–range data transfer. Optical interfaces never became widely accepted by users and eventually disappeared. Today, modern devices have on board several radio air interfaces. In particular short–range connectivity is becoming a de facto capability in addition to cellular connectivity. Bluetooth and WLAN are the most representative short–range air interfaces present in current wireless devices. Air interfaces for very short–ranges, a few cm at the most, are also becoming popular, as it is the case of Near Field Communications (NFC) technology, used for private and secure data transfer between devices or to access local information on the spot.

Different air interfaces integrated into a mobile device play different roles and typically an air interface is used in a particular scenario or with a given type of application. Cooperation between air interfaces is used in a rather simple and direct way, for instance by allowing seamless switching between two access technologies, an approach known as vertical handover. Toggling from one air interface to another can be driven by one or more events, such as channel and network conditions, mobility, coverage and others. However, the presence of several air interfaces on board mobile devices has not yet being exploited at its full potential. Dynamic and rich cooperation between complementing air interfaces opens up countless new opportunities for wireless and mobile networks to improve performance, to use resources more efficiently and to create new ways to exploit distributed resources. Cooperative approaches involving rich collaboration between cellular and local networks are classified in this book under the generic name of mobile clouds. Figure 1.3 (right) illustrates the principle of a wireless device equipped with several air interfaces, providing multidimensional connectivity across

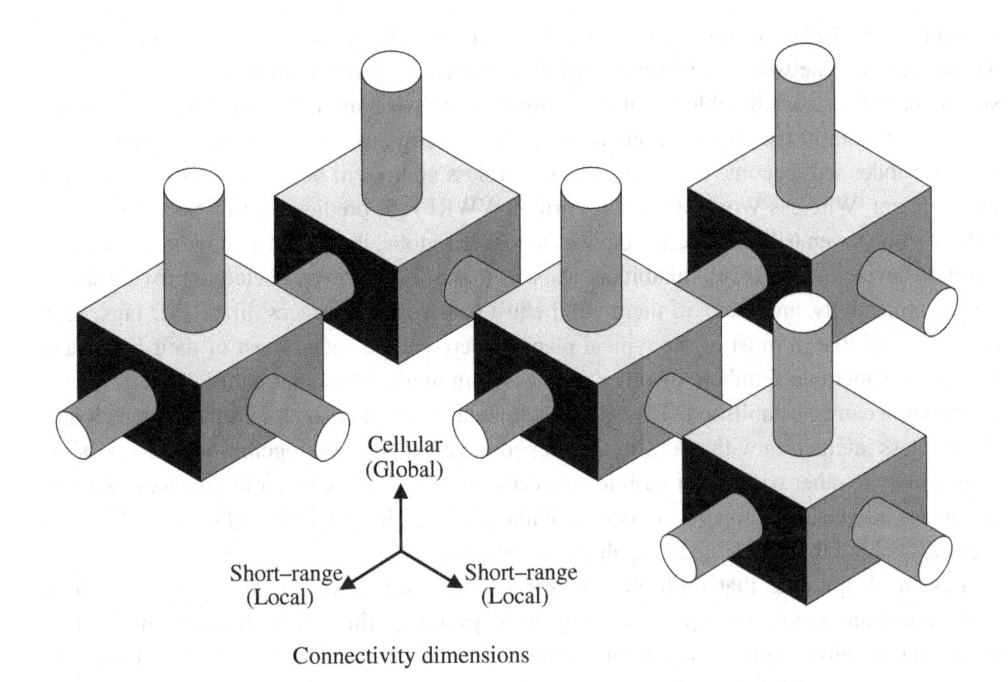

Figure 1.4 Multiple wireless connectivity (local and wide/global), the principle exploited by mobile clouds

cellular networks, local access points as well as local networks through short–range Device–to–Device communications. This represents a modern multi–air interface mobile device, providing multidimensional connectivity in the cellular (vertical) and short–range (horizontal) domains.

Figure 1.4 illustrates conceptually a mobile cloud as composed of several wireless devices (or generally speaking nodes with multiple air interfaces) that can be locally interconnected as well as can be connected to base stations or access points. This is a clear departure from the typical way to access information, or to establish connections between nodes. Indeed, the wireless network shown in Figure 1.4 is neither a conventional cellular system, nor an ad hoc network, but it combines characteristics of both. From the architecturally standpoint, the structure of Figure 1.4 retains both the centralized topology of cellular networks and the distributed topology of ad hoc networks. The composite architecture of the mobile cloud makes this approach very flexible and efficient to share resources, such as radio resources and others, like device resources. This book will discuss and investigate mobile clouds detail, considering their potentials, technical advantages, novel applications, enabling technologies, challenges and visions.

A growing trend, approximately originated at the turn of this century, is to integrate wireless communications functionalities into other devices than mobile phones. Today a great deal of portable computers, office and home appliances, cameras and cars offer wireless connectivity. Moreover, the proliferation of tiny add–on adapters using universal ports (e.g., USB–Wi–Fi, USB–Bluetooth) makes it possible to provide wireless connectivity to an even larger array of equipment. Today, in most environments where mobile users spend their time there is already a dense network of operating wireless nodes. A given mobile user is quite often surrounded by a considerable number of other wireless communications enabled devices. This is particularly true in urban environments. Certainly, as time goes by, the network of wireless nodes will become even denser, and wireless nodes will be available in virtually any environment. Wireless World Research Forum (WWRF) [5] predicts that by year 2020 there will be some seven trillion wireless devices across the globe, that is, on average one thousand wireless devices around each inhabitant. Most of these devices are expected to provide short–range connectivity, and many of them will be just passive air interfaces, like RFID tags. Still, it can be said that in most of the typical places where people spend most of their time there will be considerable numbers of other wireless communications enabled nodes with which cooperation can be established. This is a fundamental point to be exploited by mobile clouds, the wireless interaction with nodes in the immediate neighborhood together with the possible connectivity to other wireless or mobile networks. As it will be seen later in this book, the way that nodes interact and cooperate depends on many factors, including the relationship between the users behind the nodes, node capabilities and many others.

It is worth noticing that multiple wireless connectivity as shown in Figure 1.4 can be realized in many ways. Today's prevailing technology, integrating multi–standard (multi–chip) air interfaces into mobile device, is of course well suited to create the multiple connectivity approach of Figure 1.4. However, future mobile devices may have a single reconfigurable transceiver that can be readily configured on–the–fly according to a particular standard, or even as multiple air interfaces simultaneously. The upcoming LTE–A technology is an example of the developments in this direction, as it defines a single air interface supporting both cellular and Device–to–Device connectivity.

1.4 Network Evolution: The Need for Advanced Architectures

Wireless and mobile networks have steadily evolved over the past decades. This is the result of continuous R&D and huge investments by the telecoms industry and network operators, to fulfill the increasing demands and expectations of users. Improving data rates, coverage and capacity were the most important driving goals shaping this evolution. Spectral– and energy–efficiency become also important design goals for networks and mobile devices following the advent of broadband services, the rapidly growing population of users, and the massification of advanced mobile devices.

From the network architecture point of view the same topologies developed many decades ago are still in use today. Centralized access has been the key topology of cellular networks, while local networks have used either distributed or centralized topologies. These are relatively simple, well studied and widely implemented solutions. In recent years cellular network architectures have adopted simple forms of cooperation by the addition of relaying nodes between base stations and mobile devices. The architecture of cellular networks is basically deterministic, the same access topology is used regardless of the fluctuating radio environment, dynamics of the mobile devices and changing requirements of users. Local networks are by design more flexible, allowing ad hoc networking, and involving diverse types of topologies, that can be adopted depending upon particular requirements and available nodes at a given time. The stringent requirements on performance and radio resource utilization for future wireless and mobile networks call for novel networking approaches, exploiting opportunistically the fluctuating availability of network and device resources in an also changing radio environment. Conventional cellular networks lack of this flexibility, while local networks are designed with a more adaptive topology in mind. Mobile clouds, the key topic of this book, bridge cellular networks with ad hoc local networks by combining both approaches into a composite centralized–distributed topology that can react opportunistically to the changing environments and requirements. Mobile clouds offer a novel, flexible topology with an unprecedented potential not only for wireless and mobile communications but in general for exploiting opportunistically distributed resources.

1.5 Conclusion

This chapter described the state of the art and the evolution path of mobile communication systems. The number of mobile devices will increase significantly and the current mobile communication architecture will reach its limits soon. This advocates the need for mobile clouds, which will be described in the following chapters in detail.

References

[1] ITU World Telecommunication. ICT Indicators Database. http://www.itu.int/ITU-D/ict/statistics/at_glance/KeyTelecom.html, 2013.
[2] M.H. Callendar. IMT-2000 (FPLMTS) Standardization. Presented at the ITU Malaysia Seminar, Kuala Lumpur, March 1997.
[3] The UMTS Market Aspects Group. UMTS Market Forecast Study, 1997.
[4] Report by Market Intelligence Center (MIC), 2006.
[5] Wireless World Research Forum (WWRF). WWRF web page. http://www.wireless-world-research.org/.

2

Mobile Clouds: An Introduction

If you want to be incrementally better: Be competitive. If you want to be exponentially better: Be cooperative.

anonymous

In this chapter mobile clouds are defined and discussed. The mobile cloud concept is approached from different perspectives. A generic mobile cloud definition is first presented, followed by other definitions that take into account their cooperative and social aspects. The mobile cloud capabilities for sharing its distributed resources are also discussed. This chapter also discusses how cooperative and cognitive principles are exploited by mobile clouds. Finally, a classification of mobile clouds is provided, including a discussion on the main types of cooperation arising in different types of mobile clouds and their associated scenarios.

2.1 Introduction

In recent years the word cloud has become ubiquitous in many areas of computation and networking technology. Cloud computing, cloud storage/access, cloud gaming, clouds of clouds (intercloud) are typical examples of cloud technology. The word cloud is an abstraction for a system consisting of interconnected distributed resources. These resources are shared within the cloud for a given purpose, service provision being the most common objective of cloud applications. The usual services provided by a cloud include enhanced infrastructure and

Mobile Clouds: Exploiting Distributed Resources in Wireless, Mobile and Social Networks, First Edition.
Frank H.P. Fitzek and Marcos D. Katz.

processing power capabilities, ubiquitous access to information, distributed storage, security, software, testing platforms and others. The resources of the cloud can be both physical, such as hardware, as well as intangible like software. Cloud computing, possibly the most widespread cloud approach, is understood as the provision of computing services by one or more processing nodes, such a computers. The location of the nodes delivering the service with respect to the service requesting node is irrelevant, they can be practically anywhere, in remote sites, close to the requesting node, or distributed across large areas. The concept of mobile cloud computing has also been introduced, including also mobile devices as nodes. Typically a mobile user requests from his mobile device a service from a cloud computing platform. This cloud, made of nodes with enhanced capabilities (e.g., more processing power, storage capacity) executes a number of tasks related to the service and returns the requested service to the original mobile node. In this chapter the concept of mobile cloud is defined from a broader angle, encompassing mobile cloud computing as a particular case. Indeed, as it will be seen, mobile cloud computing is just one of the several applications of mobile clouds. One fundamental issue to be recognized when introducing the concept of mobile cloud is to approach mobile devices, the nodes of the mobile cloud, from a much wider perspective than normally. Indeed, in addition to providing wireless connectivity, mobile devices can be seen as featuring an array of different functionalities and capabilities such as processing power, mass memory, sensors, actuators, multiple air interfaces and others. Later in this chapter different resources will be defined in the context of mobile clouds and mobile devices.

There is a clear evolution path from current cloud solutions to mobile clouds. Current cloud solutions are single entities typically implemented in a given physical location, such as the Amazon cloud as given in Figure 2.1(a). Mobile devices are connected to the cloud service via the network operator's core network network. But such an approach is prone to failures at many levels. In 2012 the Amazon's Elastic Compute Clouds (EC2) went down due to thunder storms in Virginia. Following this event many services were disrupted, such as Netflix and Instagram, which were out of business for several hours. As clouds are introducing several new services to a growing number of distributed customers, vulnerability is increased. Therefore redundancy concepts need to be exploited by the cloud. For instance cloud replication is currently discussed among cloud providers as a way to improve reliability. Also distributing the cloud over multiple locations is currently being considered as given in Figure 2.1(b). This brings advantages not only in terms of reliability of the cloud but also considering the legal issues in the country where the data are stored or the reduction of the delay for data retrieval. Besides horizontal distribution (local distribution) of clouds, vertical distribution of clouds is also discussed. This concept advocates moving the distributed cloud more and more towards the user side, like caching by the network operator. Such concepts would decrease the delay and traffic as compared with current cloud solutions. A logical evolution of this concept paves the way to the concept of mobile clouds, where user devices form opportunistically cooperative platforms for several different purposed, as this book will be discussing.

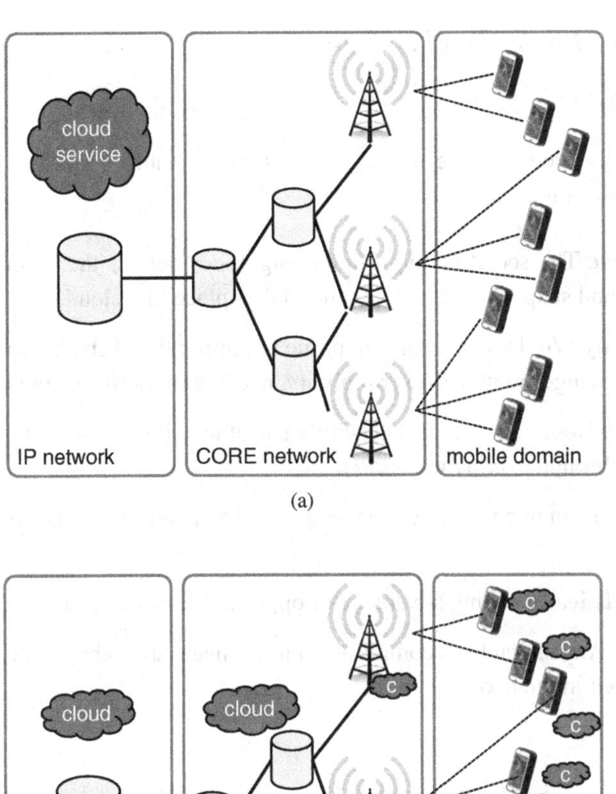

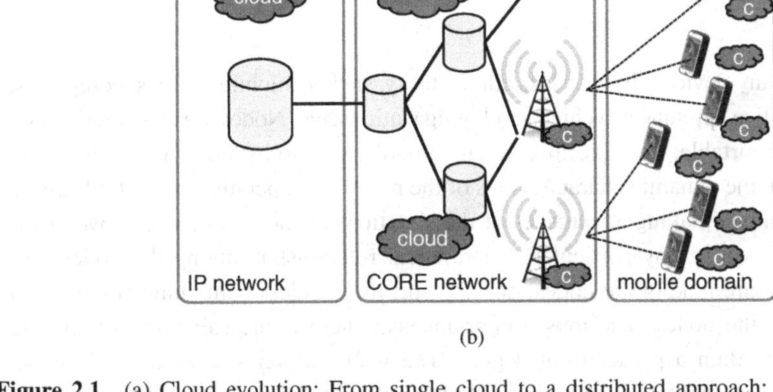

Figure 2.1 (a) Cloud evolution: From single cloud to a distributed approach: Single cloud in the backbone of the network; (b) Cloud evolution: From single cloud to a distributed approach: Distributed cloud over the whole network.

2.2 Mobile Cloud Definitions

In this section the concept of mobile cloud is defined. As mobile clouds can be approached from different perspectives, several complementary definitions will be provided. First, a generic mobile cloud definition is given, followed by a number of additional complementing definitions highlighting particular aspects of the cloud.

2.2.1 *Generic Mobile Cloud Definition*

Definition 2.1

A mobile cloud is a cooperative arrangement of dynamically connected nodes sharing opportunistically resources.

Cooperative The social relationship among nodes defines the willingness to cooperate and shape the way cooperation takes place in a cloud.

Dynamically Wireless channels are prone to temporal and spatial fluctuations as well as changes in nodes (user mobility, nodes joining in, leaving out).

Connected Nodes are connected with each other directly (peer–to–peer) or logically (through overlay networks).

Nodes Any form of communication device with capabilities to connect to each other.

Opportunistically Taking advantage of opportunities as they arise.

Resources Any sharable or composable entity/means available in the network or embedded in the nodes.

A node could be any device with wireless connectivity, such as mobile devices, computers, tablets, home and office appliances, vehicles, relaying stations, etc. Nodes are characterized by their mobility (e.g., portable, movable) but mobile clouds can also include fixed (stationary) nodes as well. Given the dynamic characteristics of the nodes, the operation of a mobile cloud needs to be opportunistic, taking advantages of the situations as they arise. The above broad definition does not assume any particular geographical relationship among the nodes, nor any explicit architecture linking the nodes, nor any specific wireless communication system technology used by the nodes. The most important issue here is the capability of sharing resources aiming to attain a predetermined goal. The goal can defined for a single node, multiple nodes, or eventually, for the whole cloud. Following this general definition, the mobile cloud can be realized over any area or region, from a local cloud to a truly widely deployed cloud. However this book will be mostly focused on a particular type of mobile cloud, where a number of constraints will make them more attractive from the standpoint of the cloud's practical implementation, as well as its achievable performance and efficiency. Indeed, most of the interest here will be given to local clouds, where nodes are close by, that is, nodes can be wirelessly connected through short–range links (i.e., typically several tens of meters at the most). Moreover, as available on an increasing number of wireless devices today, nodes can also have additional wireless connectivity, e.g., being able to be connected to access points or the overlay cellular network, through base stations. From now one, mobile clouds are defined in a more limited but practical sense, taking into consideration more concrete devices and their constraints.

2.2.2 *Mobile Cloud Definition – Cooperative Cloud*

Nodes can interact with each other through short–range links as well as being connected to the overlay cellular network or access point, even simultaneously. The mobile cloud has a composite or hybrid architecture, based on a distributed–centralized topology, combining characteristics of both ad hoc and cellular networks. The mobile cloud exhibits redundancy or over–connectivity, as in principle each mobile device can establish connections to its surrounding nodes and associated base station or access point.

Definition 2.2

A mobile cloud is a cooperative arrangement of closely located wireless devices, each of which can also be connected to other networks through access points or base stations.

Figure 2.2 illustrates the concept of mobile cloud, where mobile devices cooperate with each other locally while they can be connected to the overlay cellular network. Other names found for similar systems include wireless grids [1] and cellular–controlled short–range communications [2]. Today most people own at least one mobile device and it is also realistic to assume that in most common environments a given mobile device is always surrounded by other communication enabled devices. These devices, fixed or mobile, can belong to a

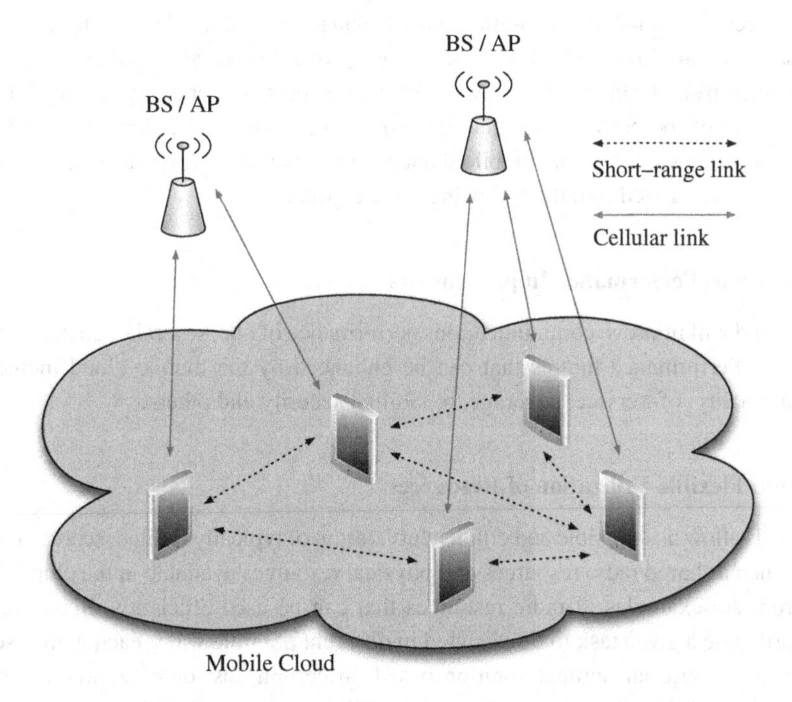

Figure 2.2 Basic distributed–centralized architecture of a mobile cloud. BS = base station; AP = access point.

given user or, in the most general case, each device has a particular owner. Mobile clouds are dynamic systems based on opportunistic interaction between mobile devices. Dynamic and opportunistic behavior is central to mobile clouds, though the degree of dynamism and requirements for opportunistic interaction will strongly depend on the type of scenario where the cloud is operating. In a public place such as an airport hall the number of devices collaborating in a cloud may fluctuate rapidly, whereas at the office or home, a mobile cloud may remain unchanged for long periods of time, as many cooperating nodes are stationary. The nodes of the mobile cloud can be heterogeneous, size–wise, capability–wise and functionality–wise. Thus, a mobile cloud can be formed by small hand-held mobile devices to bigger portable or movable devices, from simple mobile phones to advanced smart phones.

Mobile clouds are based on cooperation, and typically the user behind the device is part of the cooperative equation. Users consent to cooperate in a mobile cloud, and in principle each user may decide to what extent he will cooperate and which resources he would share with other peer users in the cloud. In some cases several nodes can be managed by a single user. Mobile clouds can also be designed to consider autonomic nodes, not necessarily controlled by users. Note that the concept of mobile cloud does not leave out the possibility of cooperative interaction embedded in the system and completely transparent to the user. In this book mobile clouds will be used as a building block for a variety of purposes, in different scenarios and for an array of possible applications. Illustrating examples of applications of mobile clouds will be provided and discussed later. At this point one may wonder, what are mobile clouds intended for? What new this concept is bringing? Why they should be used? These questions can be briefly answered with a list of capabilities and advantages of mobile clouds. These issues will be discussed in detail throughout this book. Mobile cloud performance and capabilities need to be considered from the cloud and individual users perspective. For a single user the baseline reference point will be performance and capabilities in a non–cooperative mode where the autonomic user relies only on his mobile device. The capabilities and advantages of mobile clouds can be summarized into the following three aspects:

Communications Performance Improvements

A mobile cloud will improve communications performance of one, several or all nodes making up the cloud. Performance figures that can be enhanced by the mobile cloud include data throughput, quality–of–service, coverage, reliability, security and others.

Efficient and Flexible Utilization of Resources

Mobile clouds allow also flexible and efficient utilization of typically limited, scarce and costly resources, such a shared radio resources and physical resources available in the cloud. Energy and spectrum are examples of radio resources that can be used efficiently. Since the cloud allows in principle a given task to be executed in different possible ways, each with associated costs, one can choose an optimal solution based on certain cost optimization criteria. For instance different air interfaces have different capabilities, power consumption, spectral usage requirements and associated latencies. This is particularly true with comparing short–range and cellular air interfaces.

Novel Ways of Exploiting Distributed Resources

One of the most exciting aspects of mobile clouds is the huge potential existing in the concept of exploiting opportunistically resources that belong to different nodes of the cloud. Resources can be combined in new ways, allowing the development of advanced group services and applications. The centralized–distributed topology allows optimal management of the resources. In general and due to several physical constraints, mobile devices are limited in capabilities and performance, as compared with larger communication devices. Limitations exist in many domains, including processing power, energy capacity as well as size and form–factor related restrictions. Compared with portable and desk computers, mobile devices tend to be slower to process information, sluggish in operation and with similar or slower connectivity performance. From the user perspective, a limitation could be the fact that his device is simple and therefore unable to provide high performance functionalities or to support advanced services. Mobile clouds have the potential to fight again limitations of mobile devices, and this characteristic will be further discussed and studied in this book. One of the fundamental challenges of mobile communications is the fact that all mobile devices are energy–limited portable systems, as they are battery driven. Advances in battery technology, particularly those aiming at radically increasing energy capacity per unit volume, have developed very slowly as compared with the rapidly increasing energy demands of modern devices. The availability of a limited amount of energy onboard on one hand, with the current trend of integrating more and more power–hungry functionalities on the other hand are setting one of the toughest challenges faced by manufacturers of mobile devices today. High power consumption in a small form factor results also in high amounts of dissipated heat, raising the temperature of devices to unacceptable levels. Certainly, energy limitations create also many restrictions and undesired features to the users, such as short operating times and need for frequent battery charging.

2.2.3 Mobile Cloud Definition – Resource Cloud

Depending on the scenario or situation, the interaction among distributed resources can take place in an opportunistic or deterministic manner. A mobile cloud serves as a platform holding one or more pools of resources. Resources, defined in more detail later, not only refers to radio resources but also to device (or node) resources, such as concrete physical resources, connectivity resources, apps and others.

Definition 2.3

A mobile cloud is a flexible platform for exploiting distributed resources. Resources that are wirelessly connected can be exchanged, moved, augmented and in general, combined in novel ways.

Figure 2.3 shows the concept of mobile cloud, now from a pool of resources' standpoint. Each device is assumed to carry onboard or handle a number of resources. A mobile device can be seen as a resource bank whereas the cloud as a whole is a resource pool. Since clouds are heterogeneous, the type and amount of resources can differ from node to node. Moreover,

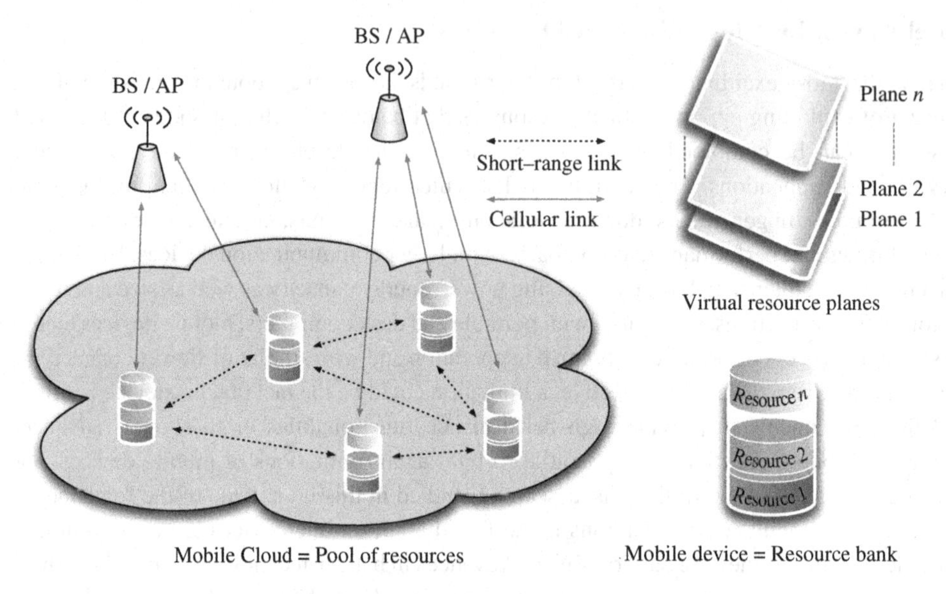

Figure 2.3 A mobile cloud as a pool of distributed resources.

even in the case all nodes are equal, the current status of resources on each device need not be equal. At a given time, a given node may have a fully charged battery while another node a nearly empty one. Equally, some of the processing power of a node could be under heavy use while at the same time the processor of another node could be idle. Distributed resources of the same kind define a particular resource plane and there could be several resource planes available. The more nodes in a cloud, the more resources are in principle available to be traded and exploited. Mobile clouds therefore can be seen as flexible platforms with an architecture that is wirelessly scalable. Examples of novel exploitation of physical resources include users combining sensors or actuators (microphones, CCD imaging devices, loudspeakers) over the cloud to achieve unique spatial processing capabilities, such a directional microphones, 3D or enhanced video/photo/audio processing, and many other possible applications. A given node can also opportunistically borrow over the air multiple functionalities from peers in the network (e.g., sensors, processing power, connectivity resources, for a particular task).

The word resource has a multidimensional meaning in the context of mobile clouds. There are multiple resources that can be exploited in and by the cloud. Figure 2.4 summarizes the basic classification of resources in mobile clouds. Any node holds a number of resources, and being for a node part of a mobile cloud means that these resources are potentially part of a pool of resources. This is a novel way of approaching resources; whenever a node is part of a mobile cloud, its resources become potentially shareable. The decision of sharing some of the available resources on the device remains ultimately with the owner of the device. The following are the most important mobile cloud resources:

(a) **Radio resources:** These fundamental resources, time, frequency (spectrum), space and energy, are closely related with the communications capabilities and performance of the

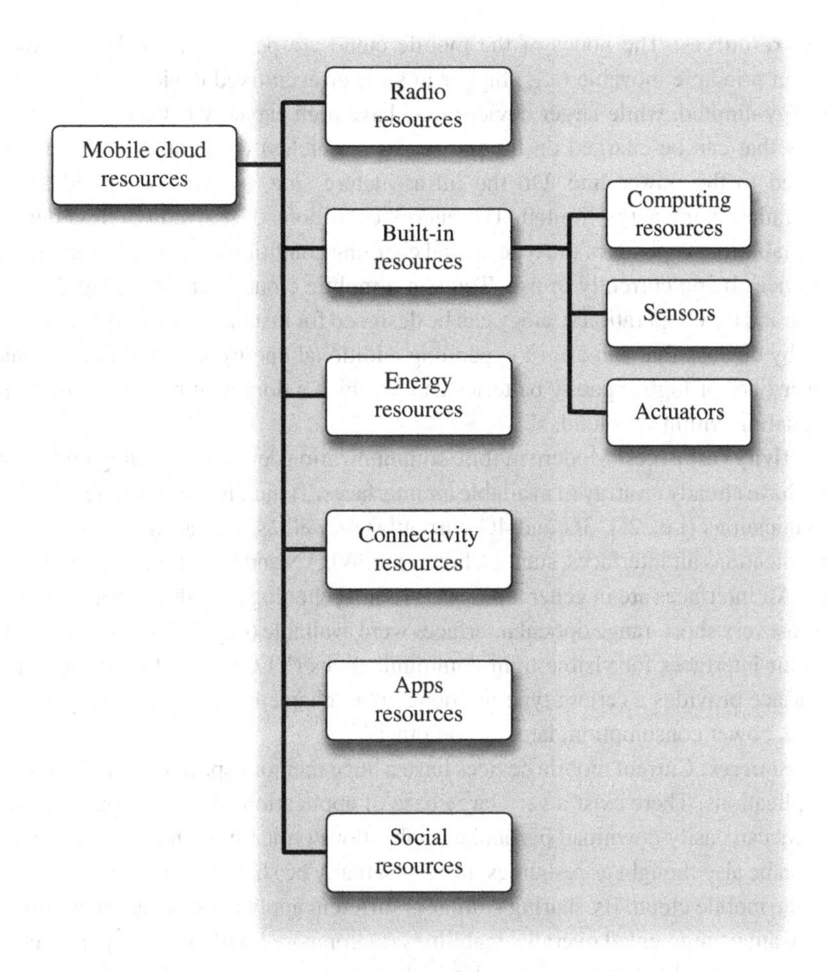

Figure 2.4 Classification of resources in mobile clouds.

cloud at both link and network level. Radio resources are the most important resources to take into account when devising the cooperative strategies to be implemented by the cloud.

(b) **Built–in resources:** These are physical resources (e.g., tangible) available on each node of the mobile cloud. They can be divided into the following types:

1. Computational resources: Mostly represented by processing power available on the node, like CPU, DSP, mass memory and graphical processor chips.
2. Sensors: Any type of sensing component used onboard of the mobile devices to sense its own status as well as the surrounding environment. Possible parameters that can be sensed include temperature, device position, device location, microphones, imaging sensors, keyboards, pollution, radiation, pollen and others.
3. Actuators: Counterparts of sensors, the actuators include loudspeakers, displays, lighting sources and any other possible physical component that can be commanded to produce a perceptible and controlled output.

(c) **Energy resources:** The nodes of the mobile cloud are portable (mobile), but they can also be in principle movable (e.g., bigger in size) or even fixed devices. Smaller devices are energy–limited, while larger devices may have high capacity batteries (e.g., laptops), batteries that can be charged on the move (e.g., vehicles) or eventually devices can be connected to the power line. On the infrastructure side, systems are typically power limited rather than energy limited. The energy conditions of a given wireless device is a time variable that depends of the type, actual charging conditions and type of functionalities and services being currently in use. Thus, in a mobile cloud energy is a highly valuable resource and the cooperation strategy can be designed for instance to favor devices with less energy by devices that can afford expending additional energy. Clearly nodes connected to power lines or high capacity batteries can be given a dominant role in handling traffic and signaling within the cloud.

(d) **Connectivity resources:** Modern mobile communication devices, including middle–range models have already an array of available air interfaces. Typically there are a few for cellular communications (i.e., 2G, 3G and 4G currently) as well as several different short–range communications air interfaces, such a Bluethooth, WLAN and Near Field Communications (NFC). Air interfaces are in general based on radio technology but this is not the only case. In the past very short–range optical interfaces were available (e.g., IrDA) while in the future optical air interfaces for visible light communications (VLC) may also be present. Each air interface provides a certain type of connectivity characterized by topology, supported data rate, power consumption, latency and others.

(e) **Apps resources:** Current mobile devices have a huge memory space intended for user data and applications. There exist a very large base of applications for each operating system, and users can easily download particular applications typically with low or no cost at all. Apps can be also thought as resources that can actually be shared, combined and expanded across the mobile cloud. By sharing similar or different applications over the mobile cloud one may attain augmented overall capabilities (as compared with non–cooperatives cases) that can be enjoyed by some or all nodes in the cloud.

(f) **Social resources:** A mobile device is normally owned by a user who has full control on the operating mode and other device settings. Ultimately it is up to the user to decide whether his device will join or not a given mobile cloud and, when in a cloud, what resources will be shared. In this sense knowing or predicting user behavior is important in order to develop cooperative strategies for the cloud. Furthermore, knowing social or group behavior is as important as knowing individual behavior. In general, one can expect that people will join a cloud because of benefits that can be attained by doing so, where benefits will be enjoyed by the user as an individual as well as by members of the cloud as a social entity.

A mobile cloud can provide connectivity channels between people, serving as a framework supporting social interaction. Thus mobile clouds can be defined from a social perspective.

2.2.4 Mobile Cloud Definition – Social Cloud

Local social interaction, taking place within the cloud, is supported by short–range links. This is an intra–cloud or small–scale social interaction. Wide social interaction (large–scale), establishing connectivity between users not in close vicinity, requires also the involvement of base station or access points. An inter–cloud social interaction takes place when connected users belong to different clouds, and a cloud-to-user interaction when one or more users are autonomic, outside of a mobile cloud.

Definition 2.4

A mobile cloud is a flexible platform for establishing mobile social networks, that is, networks where interacting users have freedom to be mobile.

Figure 2.5 shows an example of a mobile cloud seen as a platform for establishing social mobile networks. The mobile cloud provides different options to connect users to each other, each with associated costs, in terms of energy and spectral usage, achievable performance, etc. Interaction and exchange of information can occur in many different manners, such as broadcast, multicast and unicast. For instance, there are many topological choices to connect users 1 and 2 in Figure 2.5. Note that the participation of the cellular network is not ruled out, either in the conventional centralized way or by assisting a local connection, for example by authenticating or registering it. Connecting users 1 and 3, members of distant clouds, requires the use of the core network, therefore base stations and/or local access points need to be included. Certainly social connectivity to a user not part of any cloud, such as user 4 in Figure 2.5, can also be provided (cloud–to–user).

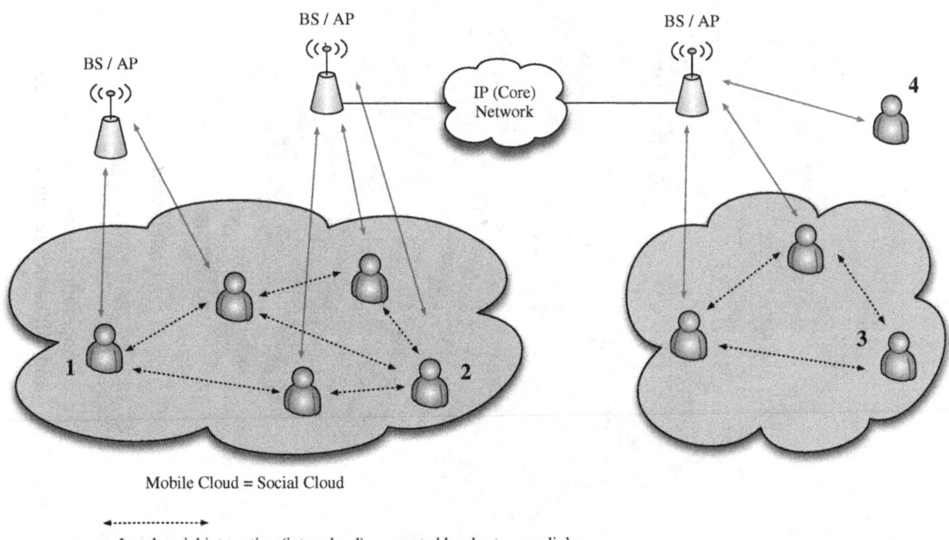

Figure 2.5 Mobile clouds as a platform for social networking.

2.3 Cooperation and Cognition in Mobile Clouds

There are many techniques in wireless and mobile communications that in one way or another are related to mobile clouds, though the proposed mobile cloud concept encompasses many well–known network technologies as particular cases. Mobile clouds are not ad hoc networks, femto cells, wireless sensor networks in their traditional definition, though mobile clouds share some of the characteristics of these technologies, at least from an architectural perspective. A mobile cloud without connections to base stations or access points is simple an ad hoc network, for instance. There are however two principles that are fundamental to mobile clouds, and are exploited as much as possible, namely cooperation and cognition. Cooperation is the most evident, and it is inherently part of the local interaction among nodes of the cloud and between nodes of the cloud and access networks. These cooperation domains are illustrated in Figure 2.6. Cognition in the mobile cloud context can be defined as a) the capability of being aware of the current conditions of the cloud (status of resources) and its surrounding environment, and b) reacting and adapting in an intelligent manner to the observed conditions. As the system and environment conditions change with time cognition is typically implemented as a continuous cycle, the cognitive cycle, as shown in Figure 2.6. Note that the observation part of the cognitive cycle is actually carried out either by sensing some resources of interest or just by direct signaling. In the former case, a particular functionality carries out the sensing operation (e.g., through a parameter estimation algorithm), as it is done in cognitive radio

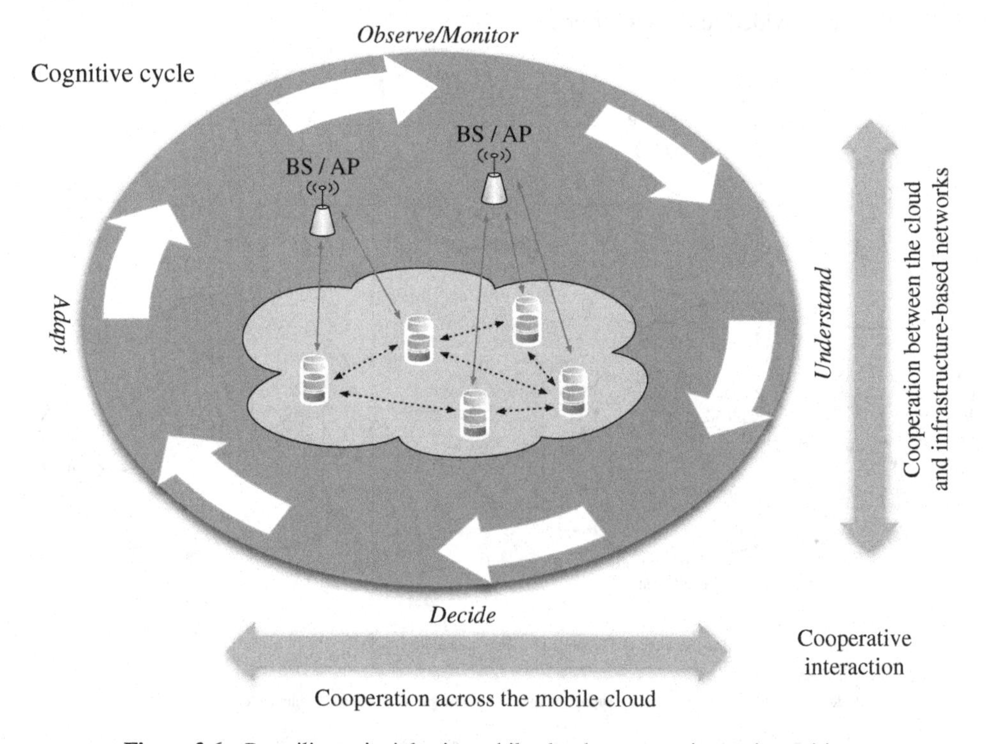

Figure 2.6 Prevailing principles in mobile clouds: cooperation and cognition.

systems, where a spectrum-sensing algorithm determines the state of the spectral domain. Direct signaling means that the current status of a given resource is just transmitted down a radio link to the node or nodes requiring that information. Such is the case for instance when a mobile device reports for instance its own measurement of battery level.

Cooperation and cognition are the main resource-trading principles exploited by the mobile cloud, whereas short–range links and access networks provided can be considered as the resource-trading domains. Though cooperation is the most essential characteristic of mobile clouds, cognition is not necessarily required when implementing a cloud, and many simple solutions are based on cooperation alone. As systems become heterogeneous the role of cognition becomes more and more relevant. Nodes in a cloud will be in general different, each characterized by its particular resources and thus taking smart decisions or implementing cooperative strategies in a cloud will greatly depend on information available about the composition of the cloud. Even in cases of a cloud with similar nodes, available resources can vary from Device–to–Device based on current and past resource usage. Cognition makes available updated information of the cloud and the environment, and this is fundamental for supporting opportunistic behavior. It is worth noticing that implementing a cognitive cycle in the cloud always consumes resources and it some cases this could be lead to prohibitively high costs (energy, spectrum, complexity, latency, etc.). A challenge for the designer is to devise solutions that are a good engineering compromise between performance, overall resource utilization and complexity. In recent years the concept of cognitive radio has been widely studied. The driving force behind this concept is the fact that radio spectrum, a limited natural resource, is on one hand becoming highly congested due to the increasing popularity of high-bandwidth services, and, on the other hand, the rather inflexible spectrum allocation schemes that have been used in most mobile communication systems. Spectrum is considered a common resource that can be shared in cognitive radio systems. Upon detection of a segment of unused spectrum the system can allocate a user to use the detected free band for his own transmission. Cognitive radio can be considered as one of the first serious attempts to exploit opportunistically common resources in a communications network. A cognitive radio system is characterized by a continuous cognitive cycle, where spectrum is sensed and based on the current observed usage; the system makes decisions on allocating unused segments of the spectral domain. As mobile clouds support a very eclectic array of resources that in principle can be shared, the concept of cognitive radio can be extended to cover other resources as well. Figure 2.7 illustrates how the cognitive cycle is implemented in a conventional cognitive radio (lower left corner) and how this concept can be extended to include other resources available in the cloud and surrounding environment. A generalization of cognitive radio, referred here to as cognitive network, can exploit in principle many resources, very different in nature. These resources are distributed across the cloud, and all of them have the potential to be common resources of the mobile cloud, as long as the users decide to share them. Users join and form a cloud for a given reason, for example expecting to gain something from cooperation, because of attractive cloud–based applications or just altruistically to help other users. Participating in a cloud means putting one or more resources on users' device into a resource pool, that is, assuming those resources to be common. Users decide which resources and to what extent resources are made common. When several nodes are owned by a particular user, it is clear that the user has full command of

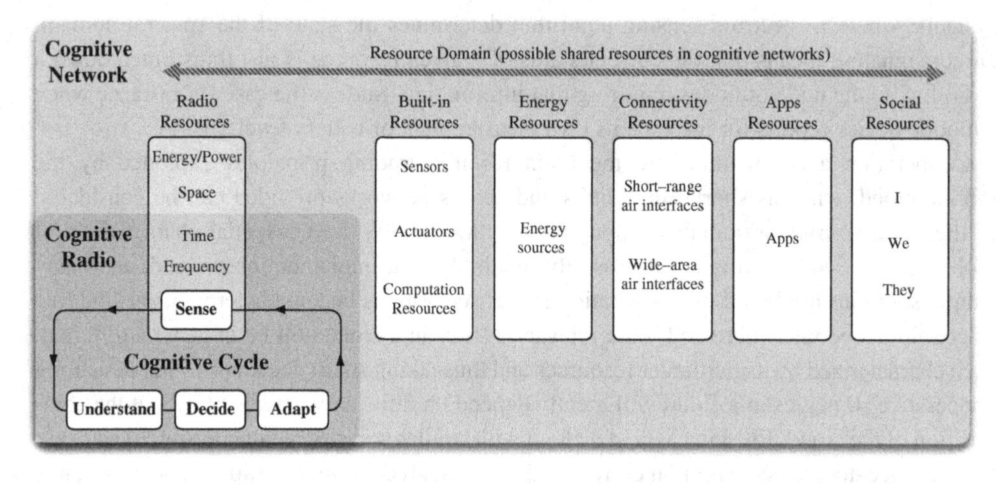

Figure 2.7 From cognitive radio to cognitive networks: the cognitive cycle allows continuous awareness on the status of common resources.

all the distributed resources on his devices. Also, the concept presented in Figure 2.7 models cases where cooperation is embedded in the system and resources are shared by design. In such case nodes work in an autonomic fashion, transparent to users.

Exploitation of contextual information is major goal when designing mobile clouds as such information will, in principle, allow the devising of high performance cooperative strategies that make efficient use of resources. Contextual information is either fixed or variable information originating either inside or outside the cloud. For instance, if information on current battery conditions of the cloud nodes is available, a cooperative strategy requiring more engagement from devices with high available battery charge, cooperatively favoring devices with low remaining charge. If a node is detected as to be moving away from the center of gravity of the cloud, this information might be indicating that the node is likely to leave the cloud, and therefore the cloud would need to be reconfigured. These two examples illustrate cases where contextual information comes from the devices themselves, or from the cloud as a whole. In addition, the cloud can make use of contextual information not residing on devices or the cloud itself, but from the environment the cloud is operating in. Knowledge on the type of environment the cloud is operating, such as public place, office or home, to name a few, gives an important hint on the type of cooperation that can be expected among the mobile devices, and this in turn is vital for deciding the most suitable cooperative strategy to be used, as will be later discussed. Some examples of contextual information:

Context information in mobile devices: type/capabilities of devices; current status or usage degree of batteries, CPU, memory and other functionalities, apps information (available, in current use), device location, speed and orientation, parameters measured by sensors onboard, owner identity and e–reputation, proximity of owner, stored user information and preferences, privileges (premium user, regular user), etc.

Context information on the mobile cloud: geometrical information of the cloud (e.g., cloud's center of gravity), relationship between users of the cloud, channel state information matrix of the cloud, etc.

Context information of the operating environment: type and characteristics of operating environment, proximity of energy sources (availability of power plugs), local policies (limitation in maximum radiated power, security issues), etc.

2.4 Mobile Cloud Classification and Associated Cooperation Approaches

The relationship between user members of a cloud defines the type of cooperation that can be expected to take place in the cloud. Three basic types of mobile clouds are defined, namely personal, private/professional and public clouds. In practice real clouds could be more complex, combining elements of these three types. However, defining these basic types is important as it will allow understanding of the nature of cooperation associated with that particular cloud. Figure 2.8 illustrates these three basic cloud types, their typical operating environments and associated prevailing cooperation. A personal mobile cloud is a cloud formed

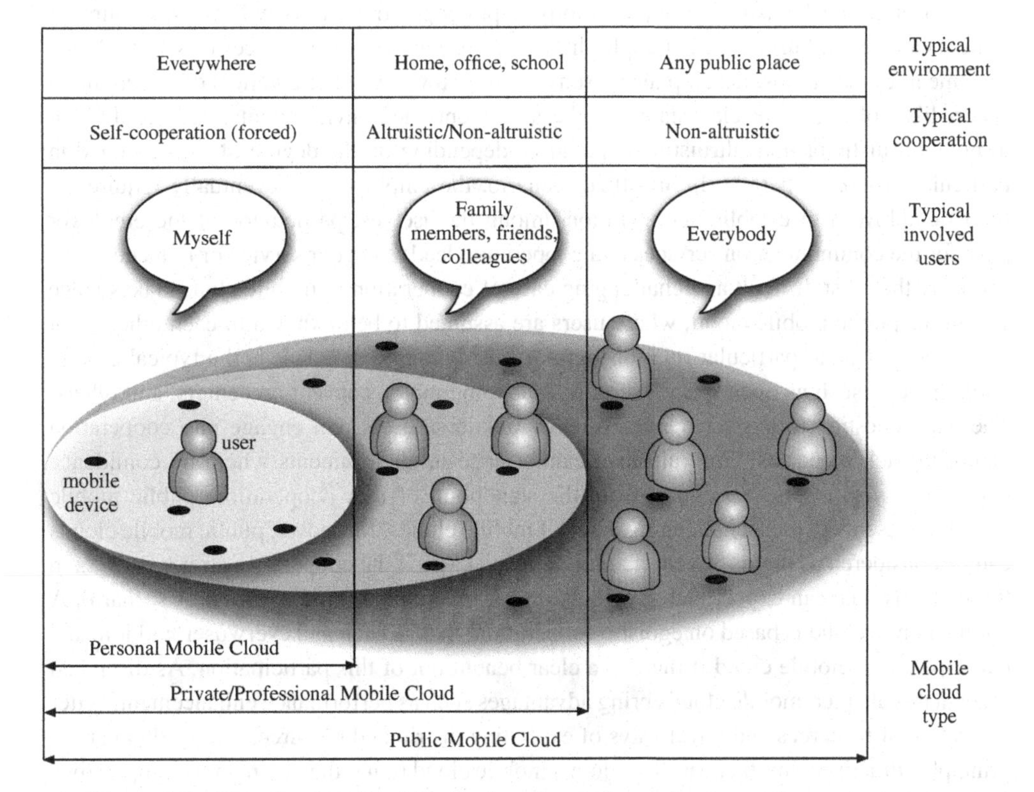

Figure 2.8 Classification of mobile clouds and typical associated operating environments.

by mobile devices or any other communications–enabled nodes all owned or administrated by the same user. As the user owns or controls all nodes, cloud resources can be exploited in any convenient manner to him. The user in this case is the only beneficiary of cooperation, and from a cloud perspective the user is forcing his devices to serve him. In this sense forced cooperation is considered as the prevailing cooperation approach in personal mobile clouds. There is no specific environment where personal mobile clouds are more likely to be formed, these clouds can be established wherever a given user and his supporting devices are. Note that, in general, forced cooperation is one of the most common cases of cooperation, occurring whenever a communication device (owned by a user) is employed as a relaying station in multi–hop systems. The relaying station gains nothing out of the cooperation; on the contrary, it does contribute with its resources (energy from battery, own time, processing power) and therefore it can be seen as being abused (forced) by the source node. Relaying networks, as used in long–term evolution (LTE) mobile communication systems, are also a typical case of forced cooperation. In a private/professional/trusted there is a closer relationship between the users behind the wireless devices. At home, for instance, one can expect that the owners of other devices are typically family members, close friends or trusted individuals. A close, familiar or affectionate relationship between users means also a relationship of confidence and trust, therefore the main approach to cooperation in a private mobile cloud is altruism. Users can share resources or help others without expecting something in return. No motivation mechanisms are required to prompt cooperation, users will naturally support or help each other. Furthermore, people sharing for instance working spaces or school floors, to name a few scenarios, develop also a sense of trust towards other dwellers of the common space, like colleagues or classmates, in the aforementioned environments. This could lead to both, altruistic or non–altruistic cooperation, depending on the degree of trust attained in particular groups of people. In an office scenario, the employer can eventually require (or force) employees to establish cooperation among themselves, particularly if the employer pays for the communication services and cooperation leads to better service or reduced costs. Probably the most difficult and challenging case of cooperation in mobile clouds takes place within the public mobile cloud, where users are assumed to be unknown to each other, or at least there exists no particular relationship nor trust among them. This is the typical case in public places, such as open spaces, airports, shopping malls, convention centers, and others. The main question to answer here is whether unknown users will engage into cooperation or not in such scenarios? Pure altruism cannot arise in environments where no confidence between players exists. No doubt, from the standpoint of user cooperation, public mobile clouds represent the most challenging cases of mobile clouds. Moreover, public mobile clouds can be considered as the most generic mobile cloud case. If the cooperative strategy works in these clouds, one can expect that it will also work in any other kind of cloud or scenario. A public mobile cloud is based on egoistic cooperation; that is, each and every user will join and participate in a mobile cloud if there is a clear benefit out of this participation. As discussed early in this chapter, mobile clouds bring advantages such as performance enhancement, better utilization of resources and novel ways of exploiting distributed resources. All of them are in principle attractive to motivate users to join a mobile cloud rather than being non–cooperative. In general, there should be incentives for a user to join a mobile cloud. Being cooperative has to pay–off as compared with being autonomous. Incentives for a user could be attaining better QoS, prolonging operating time or accessing particular cloud–based services. There exists also a financial aspect of cooperation, principally when network operators or service providers are considered. The role of the operator could be very important in mobile clouds as basic operations such a delivering of information to mobile users could be carried out consuming efficiently common resources such as spectrum. Cooperative users could be also seen as helping operators providing better services. As such, network operators could motivate cooperation by providing financial rewards to users sharing their resources, that is, opening their devices for others to use. Cloud operation could also lead to a larger number of users using a given service, benefiting the service providers, which could also devise motivating pricing policies encouraging people to access a service as a cloud.

Figure 2.8 shows the cooperation domains and typical incentive associated with the cooperation types. Certainly this is a classification for guideline purposes. Knowing the dominant or expected kind of cooperation is important to design the most appropriate cooperative strategy in the cloud.

2.5 Types of Cooperation and Incentives

In the previous section a basic classification of mobile clouds was presented. It was seen that the relationship between users defines the kind of cooperation that will potentially occur in the cloud. This section will shed more light on the types of cooperation and the typical incentives associated with the different cooperation approaches. This is important, as understanding users' behavior and their willingness to cooperate in a cloud is fundamental to design and implement a successful cooperative strategy. How costs (C) and benefits (B) of cooperation interplay can be used to model the attractiveness or appeal of a given cooperative strategy. In fact, a user's decision on whether to cooperate or not will depend basically on how he assesses this straightforward cooperative equation

$$I = B - C, \tag{2.1}$$

where I stands for the overall perceived incentive. As expected, the larger I the more motivated the user will be to cooperate, and vice versa, and systems based on user cooperation should aim at maximizing I. The benefits B can be seen as incentives known, presumed or inferred by the user, or offered to him; that is, any possible gain the user can take advantage from. The incentives or possible gains exist in many possible domains, technical, personal, social, and economic, for instance. Costs C, also spanning a multi–dimensional space, include, among others, user's resources expenditure, performance drop and risks (e.g., security and privacy issues) resulting from the collaborative participation in the cloud. Figure 2.9 illustrates the cooperation domain in mobile clouds, including altruistic, egoistic, social, forced/self–cooperation and technical approaches. Typical incentives associated with these cooperation methods are also shown. Next, each approach will be discussed and modeled using the cooperation equation introduced above.

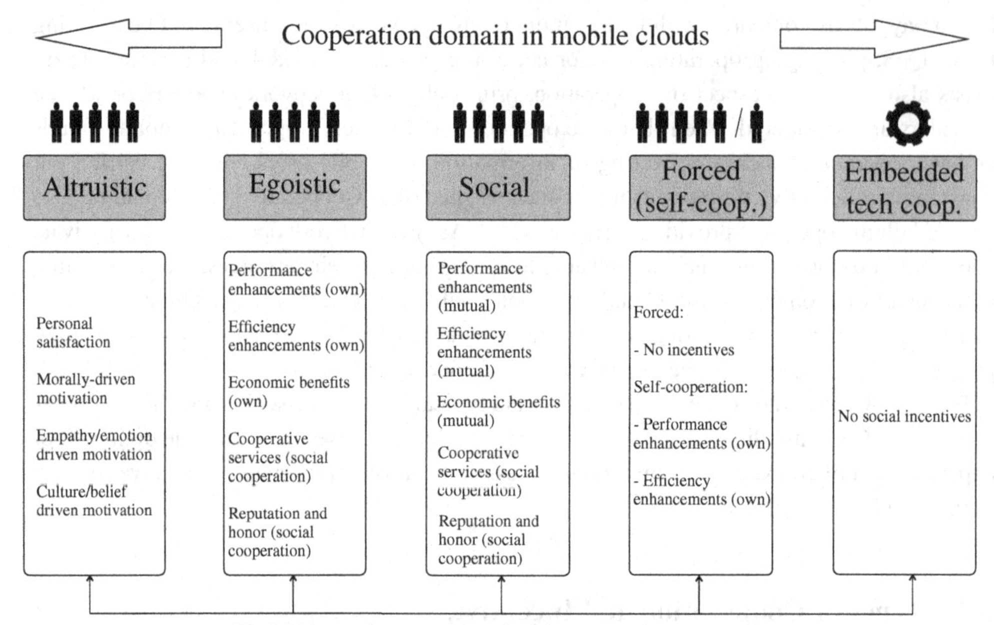

Figure 2.9 Cooperation domains and typical incentives (benefits B) in mobile clouds.

2.5.1 Forced Cooperation/Self–Cooperation

Forced cooperation takes place whenever wireless devices are requested to serve other peer devices in a forced, involuntary manner. The relationship between the requesting and requested nodes defines more precisely the kind of forced cooperation. If the nodes (devices) being forced to cooperate are on the same ownership as the requesting nodes, the forced cooperation becomes a self–cooperation, as defined later in this section. Forced cooperation could also be defined in cases where nodes of a cloud are purposely designed to take a serving role. A typical relaying station, used currently for instance in LTE networks, is a case of forced cooperation. Moreover, in an extreme case forced cooperation could also be seen just as a case of user abuse, if the imposing user makes use of illegal or unlawful methods for that purpose. Since cooperation is improperly forced, this approach can also be equated to slavery. In this case of forced cooperation $B = 0$ and therefore the perceived incentives are negative, deterring in principle cooperation. Self–cooperation, a particular case of forced cooperation, takes place typically in personal mobile clouds, where the user forms a cloud made up of wireless devices of his ownership. As such, the owner of the nodes is the only beneficiary of the collaboration. All resources available here belong to the user and the cloud will be configured to serve him. From this perspective the user does not perceive as a cost the usage of his own resources nor the possible performance drops in some devices, providing that he can obtain net benefits by engaging his devices. The possible benefits B attainable by the user by exploiting his own devices to form a personal cloud include the general items already identified: performance and resource utilization enhancements, as well as economic advantages. Moreover, the possibility of having advanced cooperative applications exploiting other resources on the personal devices can also be considered as incentives encouraging self–cooperation. The costs C perceived by the user are small or negligible, and thus, in general, one can expect that $I \gg 0$. Therefore, users are likely to form personal clouds and exploit self–cooperation, as long they can see any of the mentioned gains or advantages. Examples: Sensor networks are a typical example of forced cooperation. Indeed, by design the nodes of the network cooperate, like in the case of an embedded sensor network conveying sensorial information to a sink node. Each node contributes with its own resources, e.g., energy, connectivity and sensors, to achieve the goal of the sensor network. If all the nodes belong to the same entity (i.e., user, network or organization in this case) one can also see this forced cooperation as self–cooperation, as discussed before. In a smart grid environment, sensing reports from home appliances' instantaneous power consumption can be gathered and processed by a single device (e.g., a smart phone) and the results can be further transmitted to an energy broker using an access point or base station. Another example is a personal mobile cloud formed by sharing connectivity resources (e.g., air interfaces) of two or more mobile devices belonging to the same user, combining several data pipes to attain an increased overall data throughput.

2.5.2 Altruistic Cooperation

Altruism emerges out of a social group where members have a close relationship: in other words, people trust each other. People behind the mobile devices could be family members, friends, colleagues, classmates, for instance. From the point of view of mobile clouds, altruism can be considered an easy case of cooperation, as trust and confidence will lower obstacles hindering cooperation. An altruist is likely to take the role of donor, being ready to share his resources without expecting anything is return. Thus, rather than being concerned with their own benefits, altruistic users will evaluate both the benefits that other users can get (receptors) out of their generosity and the cost incurred to them by being cooperative. Following Hamilton's rule [3], the equation for altruistic cooperation becomes

$$I = r \cdot B - C, \tag{2.2}$$

where B is the benefit to the recipient, r is a coefficient modeling the relatedness between recipient and donor and C is the cost to the donor. The more related the users are, the higher r and the more likely it is for the donor to behave altruistically. If interacting users are not known to each other $r = 0$ and the cooperative equation becomes negative as the potential donor sees just costs involved in being cooperative.

Example: Using a mobile device to grant IP access to the laptop of a colleague is an example of altruistic sharing of a connectivity resource. Even though the cost involved includes energy from donor's battery required to share a connectivity link and the null chance for the donor to use that resource, the altruistic user will likely share his link with his colleague.

2.5.3 Egoistic Cooperation

In general, users are more likely to have an egoistic behavior rather than being altruistic. Users will cooperate if they can see a clear benefit. The easiest way to encourage selfish users to cooperate is to create situations or provide incentives such that the instantaneous benefit (B) is larger than the cost (C) for all participating users. Cooperation would happen automatically as "Real egoistic behavior is to cooperate!" [4]. This egoistic cooperation relies on cooperative technologies, and therefore it can also be simply referred to as technical cooperation. Technical cooperation driven by egoistic behavior is a type of group cooperation involving people and their decisions, and it should not be mixed up with the conventional technical cooperation which is completely transparent to users. In this book, such a purely cooperation is referred to as embedded technical cooperation. In a mobile cloud, the cooperative strategy should be developed in general assuming egoistic cooperation and therefore providing a gain to each and every user joining the cloud. As r in Equation 2.2 is small, it is clear that the users need to see a significant profit in order for them to decide to join the cloud. From a standpoint of establishing a new mobile cloud, egoistic cooperation is perhaps the most demanding cooperative approach. How to motivate selfish users to cooperate? How potential members of a cloud can be encouraged to join or start forming a cloud? How users can be made aware of the possible gains of a mobile cloud? The answers to these questions depend very much on the types of benefits offered by the mobile cloud. In general the incentives to join a cloud should come from the cooperative services or applications developed precisely for clouds. The services or applications ultimately rely on the capabilities offered by the cloud as being a flexible platform for sharing resources and also, as improving performance and resource utilization efficiency. Incentive mechanisms can be implemented in order to motivate users to perceive the benefits B as being as large as possible, regardless of the interaction with unknown peers. Note that reciprocity (e.g., tit for tat), one of the most simple and successful cooperative strategies [5, 6], cannot easily be implemented in a mobile environment. As will be discussed later in Chapter 8, there is not a unique way of perceiving a given pay–off from a given cooperative action, and understanding this is essential in order to successfully motivate selfish users to cooperate. Example: An airport is a typical example of a public place where unknown people can potentially interact collaboratively with each other. The simplest way to encourage others to start cooperation is broadcasting locally information of the available services being used and clouds already formed. This can be done through cooperative service discovery mechanisms. The broadcasted information will encourage potential users to join services being locally used. Users may end up joining services not originally considered, a fact that will benefit service providers. In addition to the running services' announcements, the broadcasted service discovery information should include critical information about the benefits of joining the cloud. This information could be for instance pricing, where users could see the cost difference between being non–cooperative and being a member of a cloud. Other achievable benefits that can be announced include approximate saving in battery energy, enhanced quality of service (QoS) or quality of experience (QoE) and others, depending on the original purpose the mobile cloud was formed for. The first user to join a given service designed to be cooperative may start promoting that service, through service discovery techniques.

Based on the broadcasted service discovery information, other interested users may join the service.

2.5.4 *Social Cooperation*

Social cooperation is referred to as the collaborative interaction between users that is mainly motivated by its social impact. By cooperating a given user can a) enhance his social status or prestige; b) support the establishing of social applications or services that are particularly designed for multiple cooperative users. The main incentives for cooperation can be seen as personal and social. A cooperative user will perceive his participation as an increased personal satisfaction, that, importantly, can be concretely measured. The more cooperative a user is, the higher will be his social reputation (or e–reputation). The social reputation of a given user, group of users or an organization can be seen as a measure of certain personal and social merits. What is important is that the measure of social reputation can be readily available and visible on the social networks of the cooperative aforementioned cooperative parties. Such a social visibility works as an effective incentive for cooperation. Note that reputation points can be used in many innovative ways, not just as a personal/group satisfaction meter. Reputation points can be traded for other resources or for financial rewards from other users, groups, organizations as well as network and service providers. Example: The more a particular user opens his device for others to use, e.g., as a repeater in the simplest case, the higher will be his cumulative reputation number posted on the user's Facebook profile or any social network. The same holds for a given organization and its members that are cooperative towards other users (members or not). Being socially cooperative could lead not only to enhancing social reputation of an individual or social group but also to obtaining any of the possible gains of mobile clouds. Social rewards can be seen as earned resources that can used to acquire services or they can eventually be exchanged by other resources.

2.5.5 *Embedded Technical Cooperation*

Embedded technical cooperation refers to the cooperative techniques that are embedded into the system and completely transparent to the users. A great deal of fundamental and advanced cooperative techniques exist, including relaying techniques, cooperative coding, cooperative antenna techniques (e.g., distributed MIMO), network coding. Note that such techniques can be, and are, used also for cooperation involving users' decisions, as in many cases of mobile clouds discussed here. This book is mostly focused on social aspects of cooperation and therefore pure embedded technical cooperation as such is not considered here. Reference [4] includes a comprehensive introduction to embedded technical cooperation approaches.

2.6 Conclusion

In this chapter we gave the definition of the mobile cloud in terms of technology and social implications. The evolution path from cloudless to cloud-based communication has been given

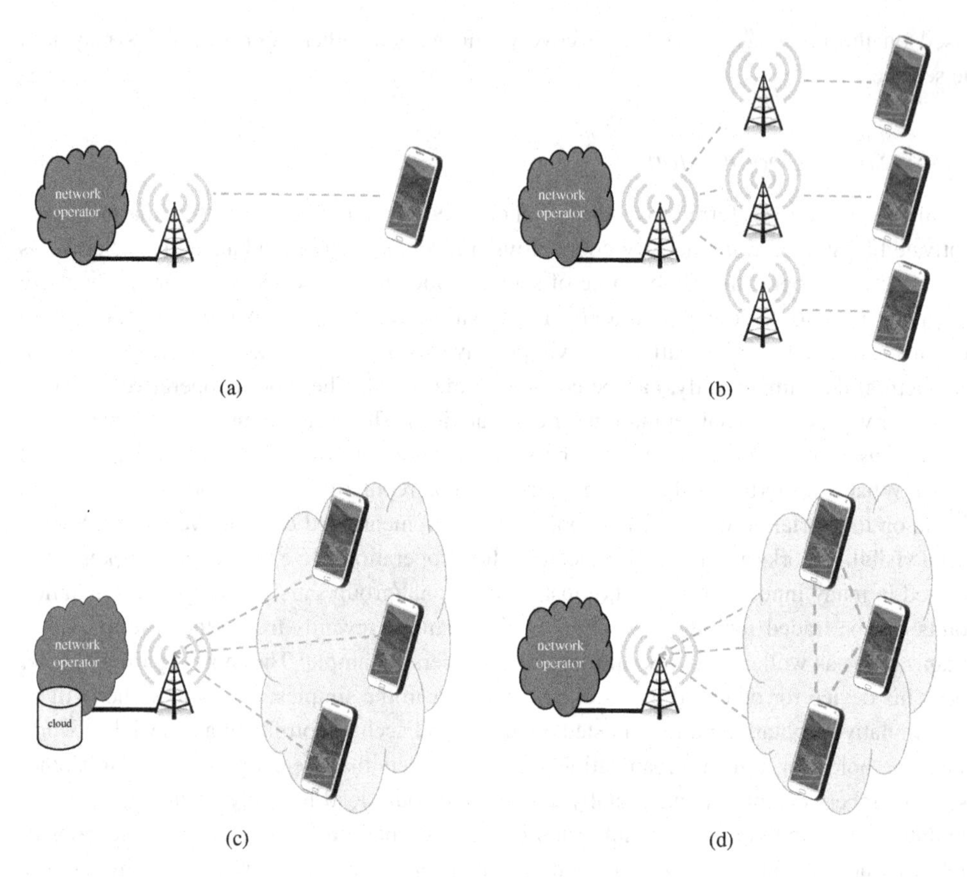

Figure 2.10 Evolution path in mobile networks from point–to–point to mobile clouds. (a) State of the art: Cellular communication dominated by point–to–point communication; (b) State of the art: Relaying communication by the network operator for coverage extension or cell off–loading; (c) Mobile cloud communication with centralized cloud infrastructure; (d) Mobile cloud with direct communication among the users and overlay communication.

and underlined that the cloud is not necessarily only based in the backbone but will converge towards the mobile device. Figure 2.10 shows the evolution in more detail: Figure 2.10(a) shows once again the point-to-point communication over a wireless link without any notion of cloud. This topology serves as a reference point for the state of the art. Figure 2.10(b) shows the introduction of communication relays in order to increase the communication coverage. The relays are most likely owned by the network operator and the cooperation is forced (see Chapter 8 for a more detailed explanation of forced cooperation). Figure 2.10(c) and Figure 2.10(d) show the mobile cloud concept as used throughout this book. In Figure 2.10(c) the mobile devices are connected via a cloud server that is realized somewhere in the Internet. The position of the three mobiles is not important as long as they are connected to the Internet. Figure 2.10(d) shows the mobile cloud when the mobile devices are in close proximity and wireless communication links can be established in a direct fashion with or even without the help of the overlay network.

References

[1] F.H.P. Fitzek, M.V. Pedersen and M. Katz. A Scalable Cooperative Wireless Grid Architecture and Associated Services for Future Communications. In *European Wireless 2007*, Paris, France, April 2007.

[2] F.H.P. Fitzek, M. Katz and Q. Zhang. Cellular Controlled Short-Range Communication for Cooperative P2P Networking. In *Wireless World Research Forum (WWRF) 17*, volume WG 5, Heidelberg, Germany, November 2006. WWRF.

[3] W.D. Hamilton. The Evolution of Altruistic Behavior. *The American Naturalist*, 97:354–356, 1963.

[4] F.H.P. Fitzek and M. Katz, editors. *Cooperation in Wireless Networks: Principles and Applications – Real Egoistic Behavior is to Cooperate!* ISBN 1-4020-4710-X. Springer, April 2006.

[5] R. Axelrod. *The Evolution of Cooperation*. basic Books, 1984.

[6] R. Axelrod and W.D. Hamilton. The Evolution of Cooperation. *Science*, 211:1390–1396, 1981.

[illegible]

References

[illegible]

3

Sharing Device Resources in Mobile Clouds

Once you have lots of different kinds of devices combining in different ways, you can not do monolithic software anymore. Each of these devices has a certain set of functions, and if we have to assume when we build them what they are going to be used for, it is not very flexible.

Bill Joy

Resources belonging to nodes of a mobile cloud can be combined in very many different ways. These distributed resources can be added up in a linear fashion to create an augmented version of the combined resources. Furthermore, resources can be combined to create particular outcomes that are unattainable solely with the resource of a single device. This chapter provides a motivating overview on the potentials of resource sharing among cooperating nodes of a mobile cloud.

3.1 Introduction

A modern mobile device encompasses an eclectic array of resources, including concrete physical resources such as sensors, actuators, batteries, complete functionalities (e.g., processing power, wireless connectivity blocks, mass memory) as well as intangible resources, including apps and information stored on the device. When collectively considering the mobile devices attached to a given wireless network such as a mobile cloud, one can also see a system of distributed resources that can be in principle wirelessly connected. Distributed resources can be tied up for a particular purpose, aiming to benefit one, several or all mobile devices and their owners. Resource sharing can be generically defined as the action of creating an active relationship among distributed resources. An active relationship means here different possible ways of combining, moving and engaging resources in the cloud, such as resource aggregation,

Mobile Clouds: Exploiting Distributed Resources in Wireless, Mobile and Social Networks, First Edition.
Frank H.P. Fitzek and Marcos D. Katz.

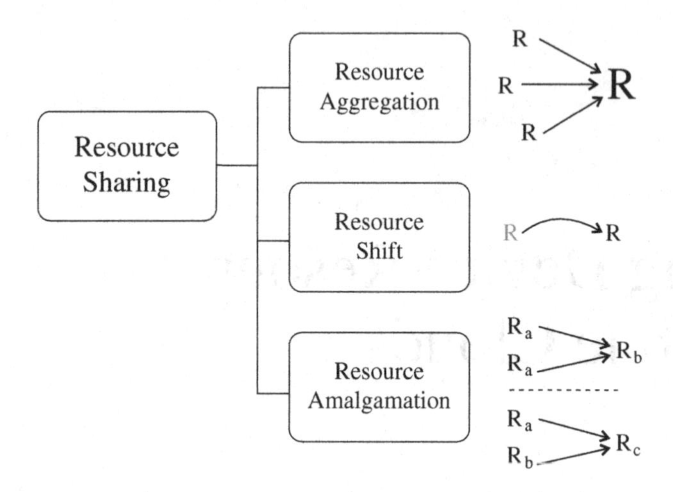

Figure 3.1 Different approaches for sharing distributed resources.

shift and amalgamation. Figure 3.1 illustrates a basic classification of resource sharing. Resource aggregation refers to putting together resources of the same kind in an additive fashion, resulting in an augmented equivalent resource. For instance, mobile screens can be puzzle–wise added up to create a screen of larger surface. Likewise, CPUs on mobile devices can be tied up wirelessly in the same fashion that cloud computing does with computers and wires, to create a more powerful processing unit based on distributed processing power. Resource shifting refers to moving resources from some devices to others, such that borrower devices can use resources of lending devices in an opportunistic manner. Thus, a basic mobile device could for instance borrow over a wireless link a resource not available onboard from a nearby advanced device. Even though resources are not physically moved to the destination mobile, the actual effect is as they were. Information gathered by a sensor in the lending node appears as sensed by the borrowing node. Such information can come from a large array of possible sensors. Resource amalgamation is defined as a combination of resources that, rather than creating an augmented version of the shared resources, projects them into a new resource space, characterized by a new dimensionality. Amalgamated resources give rise to a new type of resource that cannot be achieved with single non–amalgamated resources. Microphones and loudspeakers of multiple mobile devices can be shared in order to create spatial processing capabilities, allowing directional capture of sound as well as 3D sound effects, respectively. These extended capabilities cannot be achieved with the resources of a single device, hence the potential for creating new ways of exploiting distributed resources is an excellent incentive to encourage users to cooperate.

Resource aggregation takes place with resources of the same kind whereas resource amalgamation allows combining both homogeneous and heterogeneous resources. The conventional way of seeing a mobile device is to consider it as a device providing primarily connectivity, but also having integrated an increasing number of other functionalities and applications. We

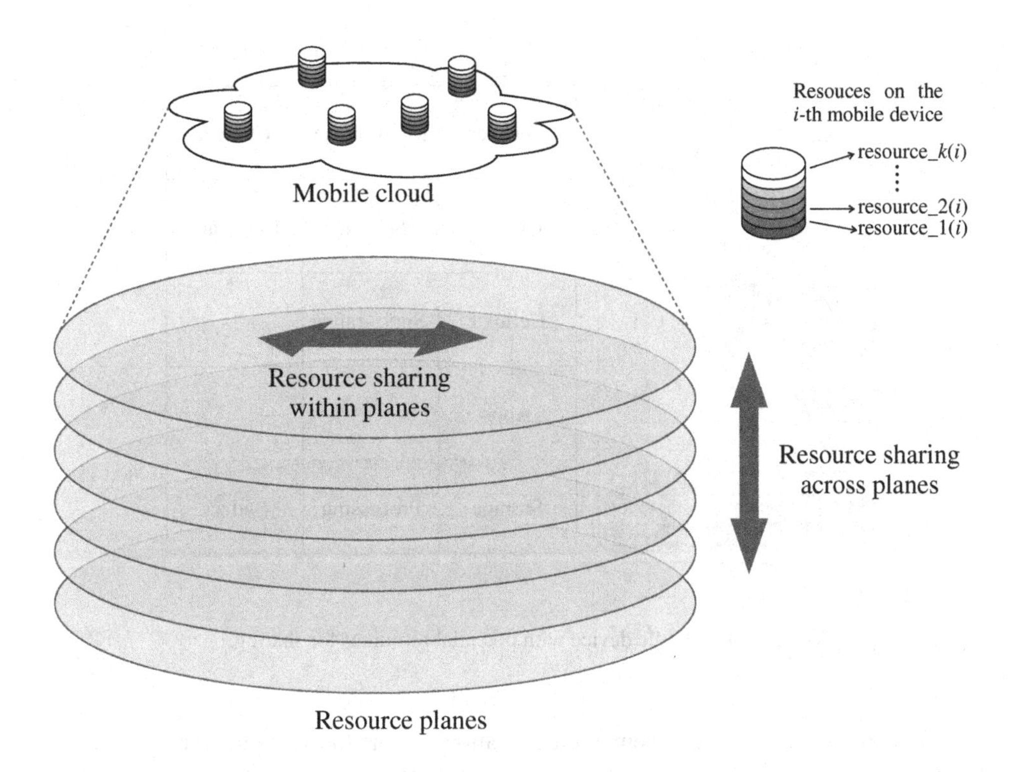

Figure 3.2 Resource sharing within and across resource planes.

can also see it as a device containing a large number of resources that potentially could be wirelessly connected with resources on other devices. One can define resources planes, so that whenever a user switches his mobile device on, he is putting the resources of his mobile device in the corresponding resource planes. Resource sharing, as described above, can take place within planes, where each plane has a particular associated type of resource, or across planes, where heterogeneous resources can be shared. Figure 3.2 illustrates the idea of resource sharing within and across planes.

3.2 Examples of Resource Sharing

Let us first identify those resources that might be subject to be shared. Figure 3.3 depicts a mobile device with typical resources onboard including speaker, microphone, imaging sensors (camera), display, sensors, keyboard, cellular air interface, short–range air interface, mass storage, processing units (CPUs, GPUs, etc), and battery. In addition to these physical resources, intangible but equally shareable resources include mobile applications (apps) as well as device–stored information. It is worth noting that device sensors embrace a very wide, eclectic and expanding array of sensing elements. In addition to the mentioned sensors,

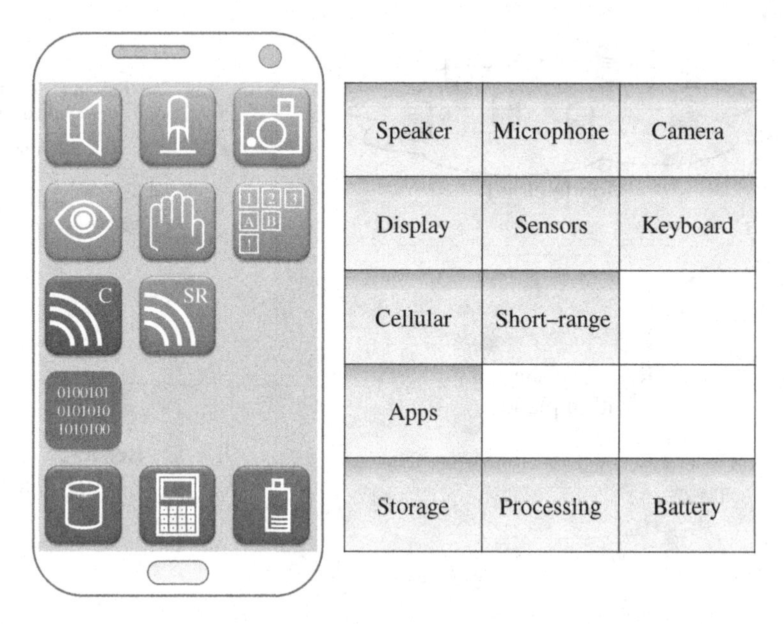

Figure 3.3 Mobile device with potential resources for sharing.

other commonly known sensorial elements are location-sensing functionalities such as GPS or cell ID, gyroscopes providing device orientation, compasses, accelerometers, and proximity sensors. Moreover, we expect that in the future more environmental sensors will be integrated into mobile devices, such as temperature, humidity, pollution, pollen and radiation. Additional sensor data can be collected by external sensors such as human heart beat sensors, blood–pressure meters, and body area networks gathering multiple physiological data from the user.

In the following, we will consider a number of concrete and motivating examples on exploiting distributed resources and related application fields. We assume that the mobile devices are connected with each other physically or in a logical manner in order to enable them for sharing. In case the mobile devices are connected physically the short–range communication will be used most likely.

3.3 Sharing Loudspeakers

When the transistor radio became popular in the 50's, young people enjoyed listening to music together in the park or at the beach; a trend that came back in the 80's with the huge *ghetto blasters* allowing youngsters literately to dance in the street. The advent of portable music players, headed by the Walkman generation, made listening to music a truly personal experience, rather than a social experience, as it was in previous decades. Nowadays, mobile phones are again being used to play music loudly, particularly among youngsters. Even though sound quality is merely good at best, people (mostly young people again) enjoy the music together rather than being isolated with headsets. Social sharing adds a new dimension to

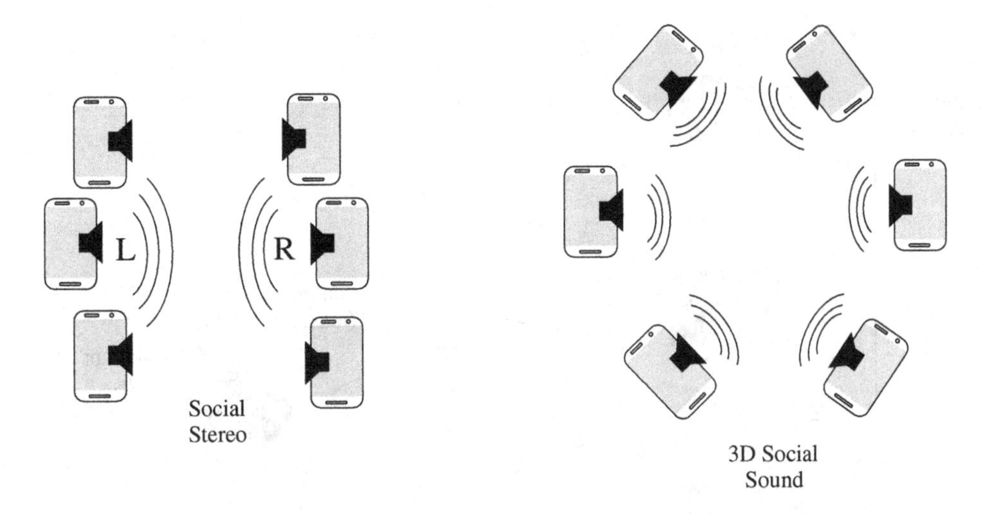

Figure 3.4 Sharing loudspeakers to create social stereo and 3D social sound.

entertainment: sharing and enjoying as a group. Instead of using only one mobile phone with one small loudspeaker, users could pool multiple mobile devices' speakers and play the music from a virtual multi–source loudspeaker, as shown in Figure 3.4. Using multiple speakers the sound experience becomes much better creating a richer audio experience, producing stereo or 3D effects, for instance. Owing to the acoustic propagation speed the distance between the devices is limited. Based on the literature [1–3] the human ear will tolerate 50ms or 100ms between the arrival of two signals from two different sources for audio or music, respectively. In [2] it is shown that the amplitude of the signal has nearly no impact. Owing to the acoustic propagation speed and the delay tolerance, the mobile devices should be in a range of 30 meters, which sounds reasonable if we consider a group of people listening to music and also keep in mind the short–range communication range. Sharing loudspeakers is a case of resource amalgamation, as the combined effect is not just an increase in the sound intensity but the creation of a new capability in the spatial domain.

3.4 Sharing Microphones

Microphones of different users could also be shared in the same way that speakers were combined to create a spatio–temporal effect. In Figure 3.5 several mobile devices record the music of a live concert. As the mobile phones are surrounded by different noise sources, their microphones will sense the same music but different noise signals. The degree of correlation of the noise picked up by the microphones will depend, among others, on the distance between theses devices. Combining the recorded streams will improve the signal–to–noise ratio (SNR) and hence the perceived music quality after the signals have been combined properly. The quality becomes better the more sources are available. Another application field is to tackle the cocktail party problem. The human being is able to focus on a single conversation at

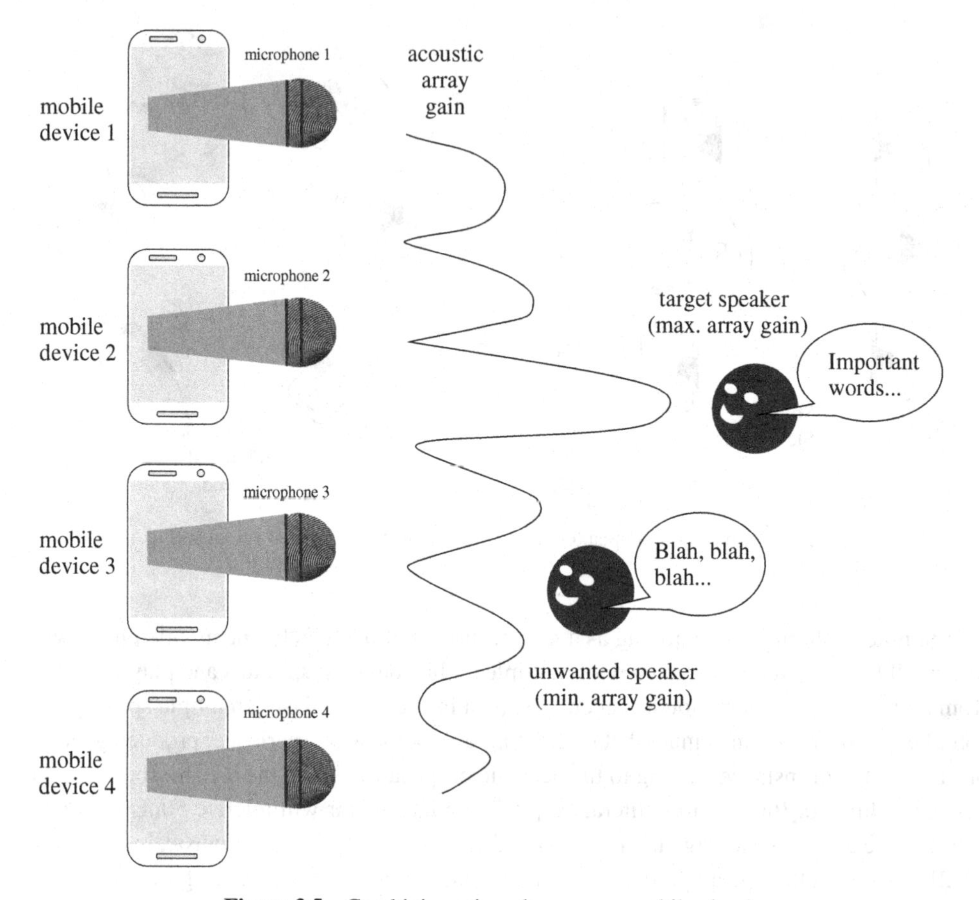

Figure 3.5 Combining microphones on a mobile cloud.

a party while other people are talking as well [4]. This is referred to as selective hearing. Some elderly people lose this capability. In order to restore the possibility to focus on the wanted discussion, all microphones on different mobile phones could be used in order to realize the separation of audio signals. The technical background to do this is called blind source separation and some very insightful examples are given in [5]. Multiple microphones can also be used to detect people's location in a room. Mobile clouds provide a platform for implementing acoustic beamforming based on social interaction, where different applications exploiting spatial filtering can be developed. Sharing microphones as described above is a case of resources amalgamation.

3.5 Sharing Image Sensors

Nearly all mobile phones have an inbuilt camera nowadays. Even entry–level devices carry basic cameras, and advanced devices can have high resolution sensors with matching top–class optics, producing high quality pictures (e.g., Nokia's 808 PureView model features a

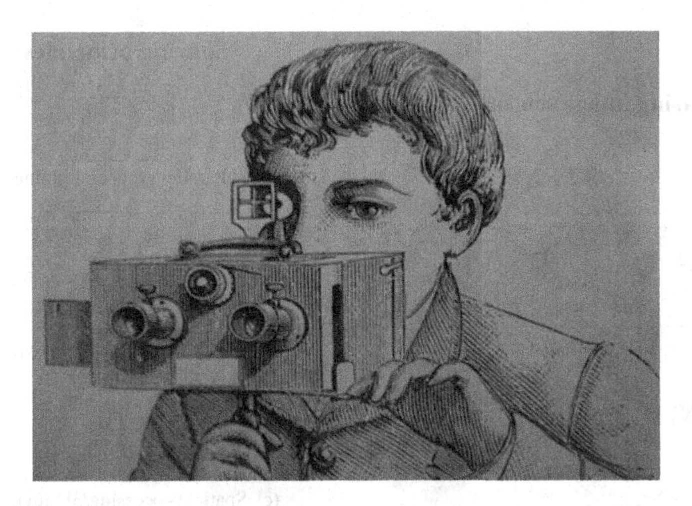

Figure 3.6 Example of an old stereographic camera. Permission to reproduce photo courtesy by Stiftung Deutsches Technikmuseum Berlin

41 megapixel sensor). But the user experience in taking pictures or video can be boosted by pooling camera sensors from multiple devices. Stereo photography was introduced right after photography itself was invented. Charles Wheatstone published his research on stereo photography in 1838 [6] and the first commercial camera was invented by Sir David Brewster in 1849. An example from the Technical Museum Berlin is given in Figure 3.6. But even later in the 1950's and 1980's the 3D pictures were very popular. But all these pictures were done by the same camera with one single click. Paul Coulton showed in [7] showed how to adopt this technology to mobile phones performing two shoots for the 3D effect.

Camera sensors can be combined in different ways, depending on how the target image is observed by the lenses of the involved cameras. Different cameras can point to the same image, as in the case of the stereographic camera of Figure 3.6, or users at a music concert pointing their cameras to the stage. Moreover, cameras can be pointed toward different directions, and hence obtaining different images that can be later merged.

In general, mobile clouds allow sharing of camera sensors in multiple ways, for a variety of purposes. Figure 3.7 depicts an example of a mobile cloud where image sensors of the nodes are pointed to a common target object. The information of sensors can be exploited according to different principles. Parallel pixel combining (Figure 3.7a) refers to concurrent scanning of the scene, where pixel–wise combination takes place, resulting in improved SNR. Figure 3.7b illustrates how resolution can be enhanced. The resolution can be increased by creating a virtual sensor with a sensing area equal to the combined areas of the shared sensors. Figure 3.7c considers the case of spatial processing by pointing at a target object from different angles. Finally, Figure 3.7d shows the case of exploiting distributed sensors in order to compensate for sensor motion. Undesired sensor motion caused, for example, by camera shakes due to hand tremor results in blurry still pictures or unpleasant unsteady videos. Since sensor motion patterns on different devices are largely uncorrelated, combining cloud sensor images can be used to compensate for the unwanted effects of motion.

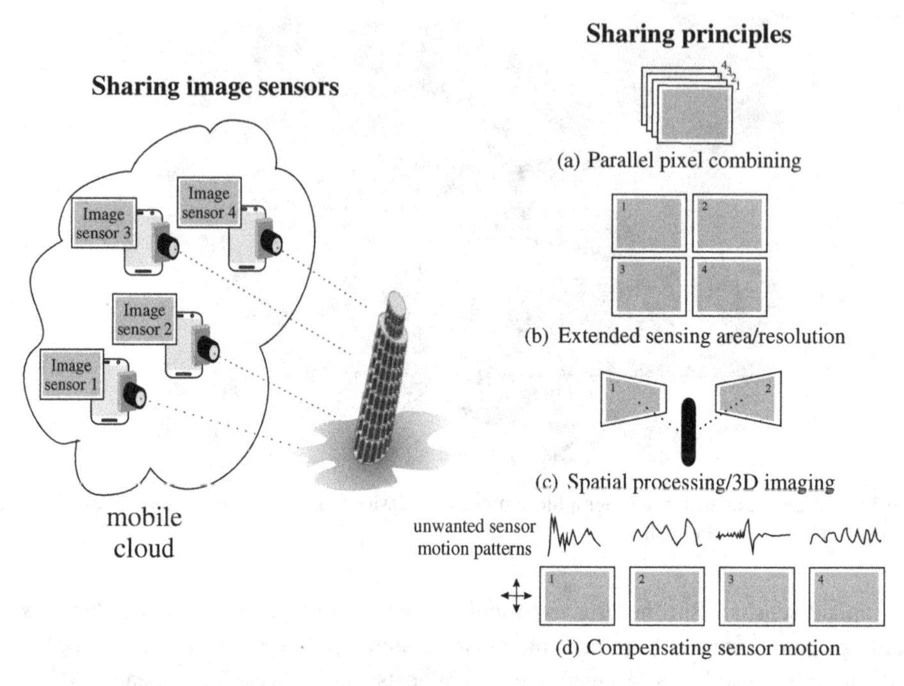

Figure 3.7 Different principles for sharing image sensors in a mobile cloud.

3.6 Sharing Displays

As users like to consume music together they might also want to watch movies or look at pictures together. In order to obtain a bigger screen, mobile users can simply combine the screens of their own mobile devices. The simplest approach is to create a single display with devices having the same form factor. The idea is not new in general as displays grids were already used in the 70's using standard TV sets. But here the idea is extended to mobile devices. Figure 3.8 illustrates the discussed approach. Four mobile phones are grouped next to each other. In order to keep it simple, the four mobile devices have the same form factor. A sending device would just need to know about the number of pooled screens and could randomly distribute the content. It would be up to the users to reorder the screen in order to get a proper picture. The combination of several screens with different form factors and different possible positions is a little bit more challenging.

Pooling camera sensors from different mobile devices pointing at different parts of a scene can also lead to interesting display aggregation solutions. Modern mobile devices are able to log not only the position and time at which a photo was taken, but also they can log the direction the camera was pointing to as well as additional settings of the camera. This allows creating pictures with higher quality or size, as illustrated in Figure 3.9, where six camera shots are combined into one. The concept is known as rotating lenses and is available on some of today's smart phones, even for single users. Combining multiple pictures of different users

Figure 3.8 Display aggregation: Four screens concatenated into one.

has more potential. First, pooling pictures of the same object from nearly the same position at the same time allows manipulating the pictures in order to get rid of blocking or undesired objects such as crossing people. Secondly, each picture taken can be linked to the same picture taken at different occasions, allowing users to see a particular scene at different times of the day, at the four seasons, etc.

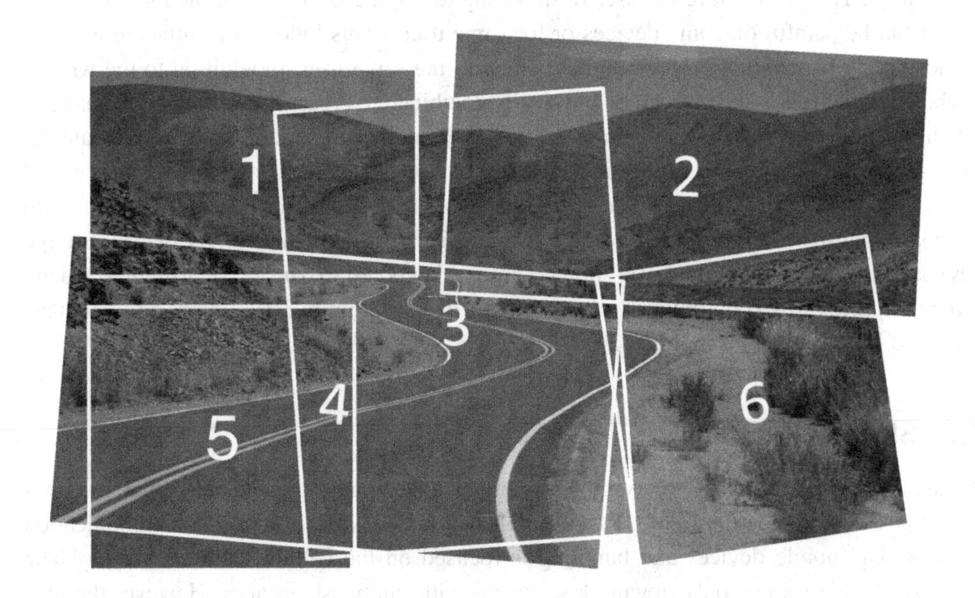

Figure 3.9 Display aggregation: Combined picture by six different cameras.

3.7 Sharing General–Purpose Sensors

Mobile phones are equipped with a large number of general–purpose sensors. Sensor data can be shared or combined for different application fields. Assuming two mobile devices that are connected via short–range technologies, a low–cost device could request the location information from a more advanced device that has GPS or other sensorial information. Even if both devices have GPS, one device might be indoors where GPS is blocked and requesting the location information from one device that is outside with perfect GPS coverage would underline the idea of sharing. Aggregation of sensor data would make sense for gathering the temperature or retrieving the traffic flow by several devices. As discussed previously, it is expected that more and more sensors will be integrated into mobile devices in the future. Users can also form mobile clouds with the purpose of gathering information about certain measurable parameters (e.g., temperature, level of pollution, etc.). This allows for creating real–time two–dimensional curves showing how certain relevant parameters are distributed. This information could be readily available to authorities and users themselves. Crowd or social sensing can take place across cities, or, in more limited spaces such as within a building.

3.8 Sharing Keyboards

The keyboard capabilities are somewhat correlated with the device size. Therefore, the idea is to use the best possible keyboard of all available devices in the mobile cloud and combine it with capabilities bound to other devices. For instance, a user could use the iPad keyboard to write a text message, which will be wirelessly transmitted to the mobile phone where it will be transmitted. This would save the user from having to use the keyboard on the mobile phone, which can be painful on some devices or for some users. This kind of resource aggregation is most likely happening on devices that belong to the same user. In addition to the writing keyboard, other touch screens can be physically combined to create a larger touching surface, for instance mobile devices can be aligned one after another in serial fashion to recreate the long keyboard of a piano or synthesizer. Devices can be placed as to simulate the distributed touch–sensitive pads of an electronic drum set. Figure 3.10 illustrates possible approaches to keyboard sharing. In Figure 3.10a a small–form device borrows from a larger device its large keyboard capabilities. Figure 3.10b shows a case of keyboard expansion by using the touch–screen displays of two contiguously placed mobile devices. Figure 3.10c shows a four–octave keyboard built out of four mobile devices.

3.9 Sharing Data Pipes

As already pointed out in several examples beforehand the bundling of the individual data rates is one possible application field. Most examples so far assume short–range communication links among mobile devices and bundling is focused on the cellular data connection. The example used considers only downlink scenarios with multicast services. However the idea is not limited only to multicast services, even though these operating scenarios will offer the largest possible gains. Unicast services will also gain by resource pooling. In [8] we have

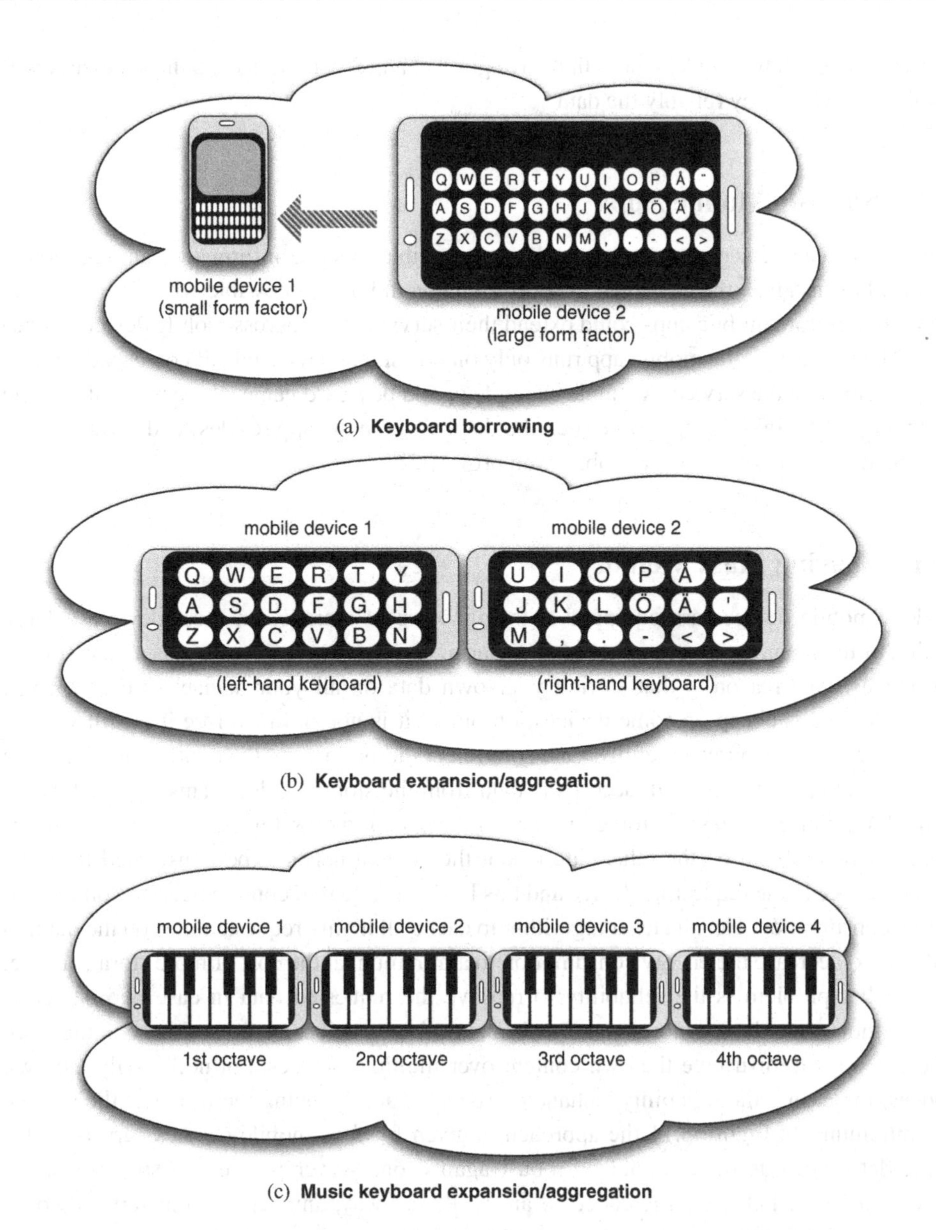

Figure 3.10 Combining several keyboards to achieve better usability.

shown that data aggregation can be used for web browsing of individual users exploiting the fact that cooperating mobile devices in the cloud are sometimes inactive and can readily help active users. From a network operator's point of view it is easier to convey information to a mobile cloud than to a single device. The cloud will ultimately deliver in the best possible way the message to the final destination, exploiting the inherent diversity of the mobile cloud. While a single device is easily prone to suffer from possible severe channel conditions, the mobile cloud will likely have some devices that have good channel conditions and those devices will be then used to convey reliably the data.

3.10 Sharing Mobile Apps

Mobile apps can be seen as enablers making it possible to share information among mobile devices. For instance, the exchange of sensor data would not be possible without the mobile apps. Furthermore mobile apps could extend their service range across mobile device boundaries. The idea is that the mobile app runs only on one single device but all connected devices can participate in the service. A simple example would be a card game where the central logic is running on the initializing device (i.e., the device where the app resides) and other devices will connect to the service via a mobile web browser.

3.11 Sharing Mass Memory

Modern mobile devices are equipped with gigabytes of storage. on most devices a large portion of the memory is not even used and sharing the free memory among users has several advantages. One reason is that storing your own data on all your devices is costly. Even though storage is cheap on some devices, on others it is not and therefore it is still wise to use this precious resource carefully. One solution could be to store the content only once on a device and other devices will access the data from the storage holder. This approach is the optimal solution in terms of storage usage. This concept works for sure on the own set of devices. The problem on the other side is that the content needs to be transmitted from the content device to the requesting device and this leads to increased communication volume and energy consumption. Here a clever algorithm to cache all highly requested files on the devices will help to decrease the energy consumption to a minimum at the cost of more storage usage. But another problem is the reliability to retrieve the requested data in case one device is down or not available any more (stolen or broken). As we will see later in this book there are different ways to distribute the own content over multiple devices, not necessarily the own devices, increasing the reliability, enhancing security, and reducing the usage of the storage to a minimum. In Figure 3.11 the approach is given for four mobile devices carrying 34% of the data. Such as setup would be robust against one server failure and still be able to retrieve the data. If the data is coded in an intelligent way, any set of three servers would regenerate the full data.

Later we will see that the performance of this approach will increase as the number of devices willing to share their resources increases. At this point we are missing the basics to give more evidence to this claim. Nevertheless the energy problem of transporting data among the mobile devices still remains and needs to be taken under consideration. Note that in some cases reliability could be the main target for sharing mass memory, and the goal would be distributing the same information across the different mass storage units of the mobile cloud. Certainly, this will come at the cost of larger memory requirements and higher energy expenditure.

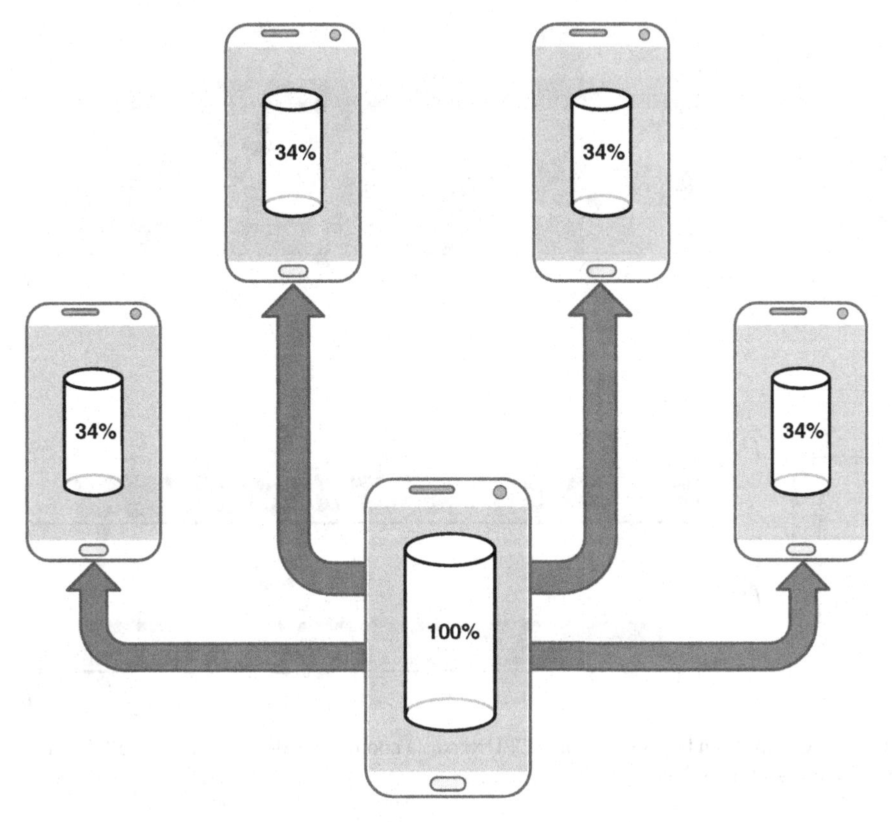

Figure 3.11 Distributed storage being robust against one server failure.

3.12 Sharing Processing Units

Today's mobile phones have multiple powerful processing units under the hood. Nevertheless for some tasks the power of the processing unit might not be enough and aggregating the processing units nearby can be a way out of this dilemma. Even if the processing power is enough it might be still beneficial to distribute the tasks due to energy reasons. As given in [9, 10] the power level of the processing unit is proportional to the square of the clock rate. Keeping the delay constraints for a certain task in mind, two mobile devices sharing one task will use only half of the power level (each one quarter) and also half of the energy (as the delay is the same for both approaches) compared with a standalone device as given in Figure 3.12. In order to exploit this feature the processing unit needs the possibility to scale the processing power. One approach support this idea is called dynamic voltage scaling (DVS). In order to be fair the drawback of the distributed approach is that the tasks need to be separable and in order to exchange the task energy will be consumed to convey the task itself and also the result back to the originating device. Furthermore the distribution will take time so that in the case of two mobile devices sharing a task will not reduce the clock rate to one half but a little bit higher than this. Therefore, it depends heavily on the scenario and the hardware used whether this

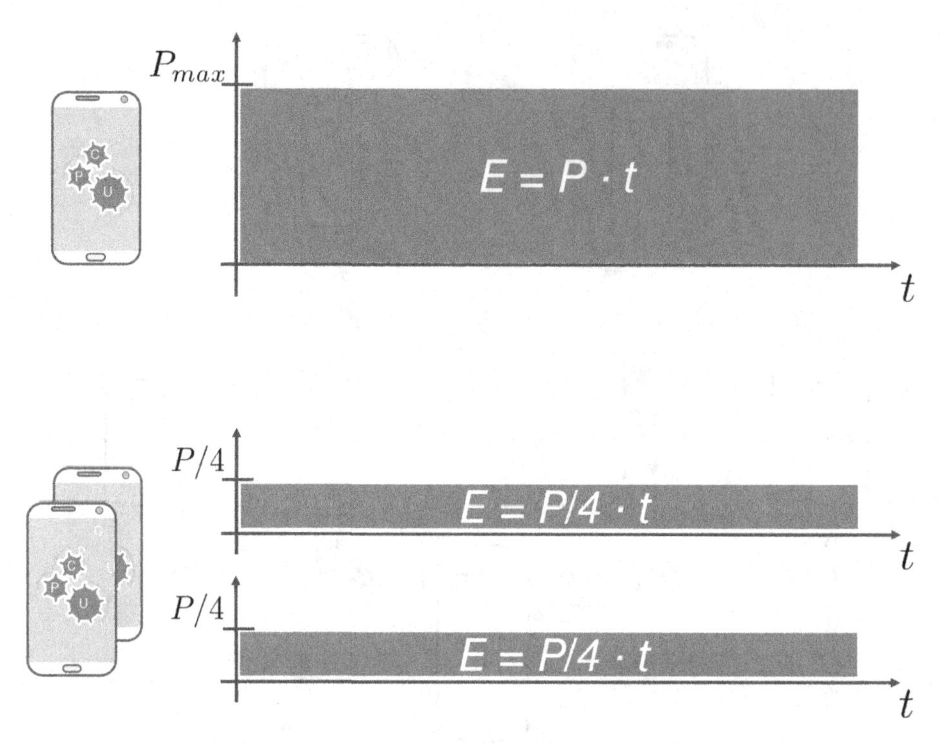

Figure 3.12 Comparison between a single CPU versus a cooperative dual core using half the energy if dynamic voltage scaling is used.

approach is beneficial or not. Nevertheless processing unit aggregation is a very interesting approach. The interested reader is referred to [9, 10]. In some cases, such as mobile gaming, mobile devices of users could carry out very similar processing, e.g., producing the same graphics for the playing devices. This can be seen as a waste of processing power resources as well as of energy. Processing units could be connected to process the information as in a distributed manner, or also, the processing task could be done by one device, sending updates, e.g., graphical screen updates, to the remaining devices.

3.13 Sharing Batteries

Battery aggregation reflects certain approaches to use the energy of the individual devices more carefully, that is more energy–efficiently. The term battery aggregation might be misleading as there are no actual Joules exchanged among the devices but tasks are distributed among mobile devices. Each mobile node of a cloud has a battery, characterized by a capacity and current charge state (remaining energy). All these energy resources are integral parts of the mobile cloud and they can be used in the most convenient way from the standpoint of the cloud. Devices with more available energy onboard can be used more generously that devices with low energy reserves. This is one of the advantages of the mobile cloud, where opportunistically users are helped by others that can afford it.

Some of the resource-sharing approaches presented here relate to other than battery concepts but still they are beneficial from a battery usage point of view. For example, the use of GPS is very energy-intensive. Therefore, a mobile phone interested in the current location could just ask via short–range the devices nearby whether they have recently retrieved a GPS position and take it from that neighbor instead of switching on the own GPS device. Another example is looking into the signalling problem. Without sharing the resources all mobile devices would have to maintain their own signalling channel. Most phones are just lying around awaiting incoming calls or data. Such a basic monitoring task costs a lot of energy. In such a scenario mobile phones could group together and install watchdogs. Watchdogs would actively monitor the ongoing signalling not only for themselves but also for their neighbors, which are not following signalling anymore saving energy in sleeping or idle mode. In case the watchdogs are aware of incoming signalling messages they will wake up the dedicated neighbor with proximity technologies (e.g., NFC) or send messages over low-energy Bluetooth.

3.14 Conclusion

In this chapter a large number of possible mobile cloud applications have been introduced. Sharing resources enables a mobile cloud to create better or even new services. In the future new resources that might be able to be shared with others will come up, but even the resources listed here are already underlining the potential of mobile clouds.

References

[1] J. Blauert. *Spatial Hearing – The Psychophysics of Human Sound Localization.* MIT Press, Cambridge, Massachusetts, 1983 and 1997.
[2] H. Haas. The Influence of a Single Echo on the Audibility of Speech. *Journal of the Audio Engineering Society*, 20(2):146–159, 1972.
[3] R.Y. Litovsky, H.S. Colburn, W.A. Yost and S.J. Guzman. The Precedence Effect. *The Journal of the Acoustical Society of America*, 106, 199.
[4] E.M. Zion Golumbic, N. Ding, S. Bickel, P. Lakatos, C.A. Schevon, G.M. McKhann, R.R. Goodman, R. Emerson, J.Z. Simon, A.D. Mehta, D. Poeppel and C.E. Schroeder. Mechanisms underlying Selective Neuronal Tracking of Attended Speech at a Cocktail Party. *Neuron*, 77(5):980–991, 2013.
[5] T.W. Lee. Web page on blind source separation. http://cnl.salk.edu/~tewon/Blind/blind_audio.html.
[6] C. Wheatstone. Contributions to the Physiology of Vision. Part the first. On some Remarkable, and hitherto Unobserved, Phenomena of binocular vision. *Philosophical Transactions of the Royal Society of London*, 128:371–394, 1838.
[7] F. Chehimi, P. Coulton and R. Edwards. Advances in 3D Graphics for Smartphones. *Information and Communication Technologies – ICTTA '06. 2nd. IEEE*, pages 99–104, 2006.
[8] G.P. Perrucci, F.H.P. Fitzek, Q. Zhang and M. Katz. Cooperative Mobile Web Browsing. *EURASIP Journal on Wireless Communications and Networking*, 2009.
[9] A. Brodlos, F.H.P. Fitzek and P. Koch. Energy Aware Computing in Cooperative Wireless Networks. In *Cooperative Networks, WirelessCom 2005*, volume 1, pages 16–21, Maui, Hawaii, USA, June 2005.
[10] A. Brodlos, F.H.P. Fitzek and P. Koch. Evaluation of Cooperative Task Computing for Energy Aware Wireless Networks. In *International Workshop on Wireless Ad-Hoc Networking (IWWAN) 2005*, London, UK, May 2005.

Part Two

Enabling Technologies for Mobile Clouds

4

Wireless Communication Technologies

The wireless telegraph is not difficult to understand. The ordinary telegraph is like a very long cat. You pull the tail in New York, and it meows in Los Angeles. The wireless is the same, only without the cat.

attributed to Albert Einstein

In this chapter we briefly describe wireless and mobile technologies, the component networks needed to form mobile clouds. Two types of communications approaches will be discussed, namely cellular and short–range technologies. The capabilities and limitations of these systems will determine the way these component networks cooperate and will ultimately determine the performance of the mobile cloud.

4.1 Introduction

There are many types of wireless technologies integrated on mobile devices that could be used to build up the mobile cloud. In the following subsections these mobile and wireless communications technologies will be presented briefly. The main focus is on cellular and short–range communications technologies. Mobile and wireless communications technologies evolved following different developing paths, sometimes referred to as the mobile or cellular path, and the wireless path, respectively. First we will discuss the cellular evolution path, introducing the key representative technologies of the mobile communications generations. Next, wireless local area network (WLAN) or Wi–Fi technologies are presented, namely different versions of IEEE802.11 and Bluetooth. Figure 4.1 depicts the supported data rates versus communication

Mobile Clouds: Exploiting Distributed Resources in Wireless, Mobile and Social Networks, First Edition.
Frank H.P. Fitzek and Marcos D. Katz.

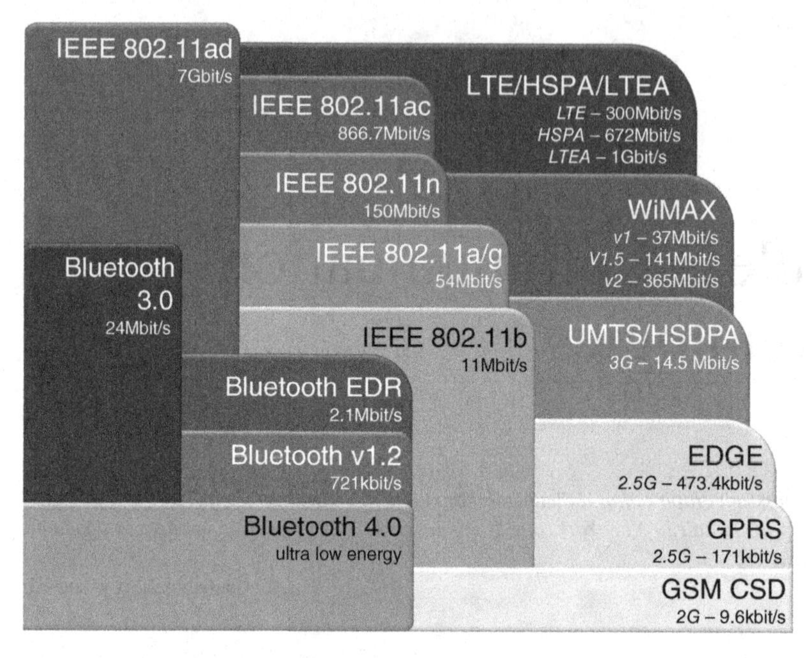

Figure 4.1 Supported data rates and typical ranges of wireless and mobile technologies.

range for different mobile and wireless communications technologies. The figure shows 2G technologies such as GSM CSD, GPRS, 3G technologies such as UMTS/HSDPA as well as 4G technologies namely WiMAX, LTE, HSPA+ and LTE advanced (LTE–A). These technologies are further explained in the following sections. Both technologies, namely cellular and short–range communication, can be used to build a mobile cloud. These technologies work in different frequency bands and can be referred to as orthogonal in their frequency usage. In the end of this chapter we also look into the future technologies where the connection to the overlay and the connection to the cooperative peers is in the same band. LTE advanced (LTE–A) is a cellular system that is supporting Device–to–Device communication by providing the needed resources for communication.

4.2 Cellular Communications Systems

Cellular technology has undergone a long and highly successful evolution path over different generations (1G to 4G) aiming at providing mobile users with increasingly higher data rate support. After the first generation (1G) of mobile communications systems, which used analog transmission schemes and had no intention to support data connections, the digital era started with the second generation (2G). The dominant technology in 2G was GSM (Global System for Mobile Communications). Originally GSM stood for Groupe Special Mobile as it was an European initiative. GSM reached global penetration while competing technologies such as IS–95, cdmaOne or cdma2000 were limited to some countries only. GSM used a bandwidth of 200kHz and the rate supported by the first data connections was 9.6kbps. This data rate

Table 4.1 GPRS data rates (kbps).

Circuit Switched type	1	2	3	4	5	6	7	8
CS1	9.05	18.10	27.15	36.20	45.25	54.30	63.35	72.20
CS2	13.40	26.80	40.20	53.60	67.00	80.40	93.80	107.20
CS3	15.60	31.20	46.80	62.40	78.00	93.60	109.20	124.80
CS4	21.40	42.80	64.20	85.60	107.00	128.40	149.80	171.20

Data rates supported by GPRS technology for different coding schemes and channel bundling.

was achieved by using one voice channel and it is referred to as circuit switched data (CSD). The data rate was ridiculously low, when seen with today's eyes. But in the 1990's voice was the main service with security and mobility. Security was already achieved by switching to digital communication and GSM technology was very expensive at that time. Mobility was needed to support customers roaming across Europe's large number of countries. As data service became more important the GSM standard allowed different techniques to increase the supported data rate. The first improvement was introduced with High Speed Circuit Switched Data (HSCSD). HSCSD improved the data rate per available channel to 14.4kbps and allowed channel bundling up to four channels, leading to a maximum data rate of 57.6kbps. The next evolution step was the introduction of General Packet Radio Service (GPRS) as an extension to CSD and HSCSD. GPRS as well as the already available HSCSD were referred to as 2.5G technology and can be seen as a first step towards the next evolution step, namely 3G networks. As in HSCSD, GPRS exploits bundling techniques to increase the data rate, but for GPRS bundling up to eight channels was allowed. Furthermore GPRS defined four different coding schemes (CS). If up to eight time slots can be bundled and four different CS are available, 32 different data rates are available as given in Table 4.1.

By bundling TDMA time slots in the GSM cell it provided realistically data rates between 40 and 50kbps (with a theoretical capacity of 171.2kbps). GPRS is part of GSM since Release 97 and onwards. In contrast to the CSD and HSCSD, which were circuit–switched, GPRS is already packet–switched and allowed the user to be *always on*. The main difference between circuit–switched and packet–switched is that in the first case a communication connection, including the communication path and the used resources, is reserved for a set of communication partners, while in the latter case the communication traffic is packetized into an information container and send possibly over different paths between communication pairs. The available GPRS bandwidth is shared among all GPRS users in the cell, where voice services are prioritized, so no quality of service (QoS) guarantees can be made. The data rate of GPRS was sufficient for email services and web browsing. Enhanced Data Rates for GSM Evolution (EDGE) or enhanced GPRS (EGPRS) increased the data rate to 473.6kbps in the early 2000's. The high data rate is achieved by extending the four modes of Gaussian Minimum Shift Keying (GMSK) modulation used in GPRS with five modes of 8–phase shift keying (8–PSK) modulation allowing 59.2kbps per channel. Later EDGE evolution was introduced, boosting the data rate to 1.6Mbps.

With the Universal Mobile Telecommunications System (UMTS) the 3rd generation (3G) of mobile communication system was introduced. The 3G technology is based on Wideband Code Division Multiple Access (W–CDMA) as the underlying standard. UMTS is standardized by 3GPP offering data rates of nearly 14Mbps using the High–Speed Downlink Packet Access (HSDPA). The boost in data rate achieved was mainly due to the bandwidth increase from 200kHz in GSM to 5MHz in UMTS.

Towards the fourth generation (4G) two competing technologies have been proposed, namely Worldwide Interoperability for Microwave Access (WiMAX) by the WiMAX Forum and Long Term Evolution (LTE) by 3GPP. Both approaches are targeting data rates of hundreds of Mbps using OFDM technology as well as advanced antenna techniques including several MIMO (Multiple Input Multiple Output) technology approaches. LTE uses 100MHz of bandwidth. At the same time an improved high–speed packet access scheme, namely HSPA+, was introduced that uses only 40MHz of bandwidth, achieving up to 672Mbps.

As can been seen in Figure 4.1 the data rate has been increased from 9.6kbps to 1Gbps over the last twenty years, a phenomenal 100,000–fold increase. The cellular technology is needed to provide connectivity on the move to each mobile device of the cloud, to connect mobile users belonging to different clouds and as well as a means for seeding information into the mobile cloud.

4.3 Short–Range Technologies

After the cellular technologies here we present the basic concepts of Bluetooth and Wi–Fi (IEEE802.11). Even though there are a large number of other short–range technologies, here we are mainly looking into those two technologies that are extensively found on most mobile devices nowadays. While Bluetooth is by far the most widely used short–range communications technology on mobile devices today (feature phones and smartphones), IEEE802.11 is breaking some ground on the more advanced mobile devices (e.g., smartphones).

4.3.1 Bluetooth

Bluetooth is a radio technology that operates in the 2.4GHz band. It is often referred to as a short–range communication technology as the range of communication is relatively small compared with cellular systems. The communication range is determined by the power class of the Bluetooth module. There exist three different Bluetooth classes, namely class 1, class 2 and class 3. Class 1 one devices can have communication ranges of up to 100m, while class 2 and class 3 devices are limited to 10m or less than one meter, respectively. Most mobile devices are class 2, while Bluetooth access points are class 1. Bluetooth systems are composed of a radio/baseband part and a software stack. Originally Bluetooth was intended as a cable replacement. The first applications of Bluetooth cases were described as connecting PCs and laptops with printers. As time evolved, a much wider range of applications were developed for Bluetooth. Bluetooth eases the process of connecting cordless peripherals such as headsets or GPS modules. Moreover, Bluetooth offers different communication profiles to define which service can be supported at a given time. Voice profiles are used for headsets being connected to a mobile phone, while the LAN profile is used for data communication between two peers for IP traffic. In the early time of Bluetooth a device could only support one of the profiles, while nowadays most if not all devices support multiple profiles. This is needed for example in case a mobile phone is connected with a headset and a PDA at the same time. Choosing the phone number on the PDA, setting up the call over the phone, and talking over the headset is only possible with multi–profile Bluetooth chip sets. Bluetooth chip sets have been advertised initially as a technology with a BOM cost (bill of materials) of five US dollars. Unfortunately, today chip sets cost around 30 dollars if bought in small numbers. Even with a larger number the five dollar threshold cannot be achieved. Bluetooth communication takes places between one master device and at least one, but maximal seven, active slave devices. All slave devices are connected with, and only with, the master device. The numbers listed here are referring to active devices. As the master is able to park a device, the master could be theoretically connected to more devices, but the number of active ongoing communication partners is restricted to seven active devices. Owing to this architecture, slaves cannot communicate directly with each other and are dependent on the master to rely the information. We would like to note that only point–to–point communication is possible, excluding thus broadcast and multicast communications for the slaves. Some Bluetooth implementations allow the master to broadcast information to all slaves at the same time. To discover other Bluetooth devices in the vicinity, each device can start a service discovery procedure. Service discovery will search for other devices in the immediate proximity and classify them into mobile phones, PCs, headset, etc. Once those devices are found, they can be paired, that is to say a procedure where devices are approved as communication partners. In the case of a large number of Bluetooth devices, the discovery process can take quite a long time. With more than ten devices around, it can take minutes to discover all neighboring devices. One Bluetooth device has the ability to support three synchronous or eight asynchronous communication channels. The synchronous channels are used for voice services mostly, while the asynchronous channels are for data communication. As we use mostly data connections in this book, we will describe those in more detail. Since Bluetooth works in the 2.4 GHz band, it exploits frequency hopping to make communication links less error prone in presence of other wireless–enabled devices using this open ISM (industrial, scientific, and medical) band. The medium access is organized in a time division multiple access (TDMA) fashion, where the channel is split into 0.625ms slots. Whenever one device transmits information to another device the successful reception of this information needs to be acknowledged in the next slot. In the case of unbalanced data transfer, such as the transmission of a photo from one device to another, one device transmits the data while the other transmits back just acknowledgments. Since acknowledgments occupy a full slot, this procedure is not very efficient. To increase the efficiency, three or five slots can be bundled by one device and are acknowledged only by one slot. Furthermore Bluetooth has the option to protect the data by forward error correction (FEC) information. Those with FEC are referred to as DM packets and those without are referred to as DH packets. Each of those packet types can used 1, 3, or 5 slots, which ends up in six different packet types namely DM1, DH1, DM3, DH3, DM5, and DH5. Whether to use DM or DH packets depends on the signal quality. DH packets offer more capacity than DM packets, but it might be the case that those packets will be retransmitted more often as they get lost and therefore are not acknowledged. The DM packets were typically used in cases the wireless medium was highly error–prone. However, recent findings show that the DH packets are more or less as robust as the DM packets. This is due to the advancements in circuit design, particularly on the enhancements attained in transmitter/receiver sensitivity. Standard Bluetooth can achieve data rates of 721kbps. Using the enhanced data rate (EDR) data rates, up to 2.1Mbps are available as given in Figure 4.1. Lately Bluetooth v3 and Bluetooth v4 have been introduced. While Bluetooth v3 targets higher data rates (up to 24Mbps), v4 is designed for ultra low energy consumption. Bluetooth v3 is not really an evolution step and not many implementations are currently available. Bluetooth v3 does the connection handshake on the own Bluetooth technology, while high data transfer is realized over IEEE802.11 links. Bluetooth v4 on the other hand is not aimed for larger data rates but it just enables the connection between devices that have long operational times such as watches. There was also a Bluetooth version planned to support very high data rate of up to 400Mbps using Ultra–Wide Band (UWB) technology. But this technology never made it into the production lines.

Forming mobile clouds with Bluetooth technology is not optimal, because of the limitations in topology. In fact, the connection setup with master and slaves makes it very vulnerable to changes in the topology (e.g., one node leaving the cell). Furthermore, the data rate have not been increased over the last ten years and the potential gain in setting up the mobile cloud becomes smaller and smaller. Without going into detail but referring to Chapter 9 the ratio of the cellular data rate and the short–range data rate has a huge impact on the performance of the mobile cloud. Therefore, the IEEE802.11 is more suited for mobile clouds as we will see in the next section.

4.3.2 IEEE 802.11

IEEE 802.11 defines a set of standards for short–range wireless communications networks, known as wireless local area network (WLAN) (see [1]). The 802.11 family is based on one medium access protocol and different physical layer implementations. At its initial stage 802.11 had three realization forms at the physical layer, namely direct sequence spreading (DS), frequency hopping (FH) and diffuse infrared (IR). As IR was limited to line-of-sight and FH at that point of time was more complex to realize than DS, all chipsets used and continue to use DS technology. FH and DS were not intended to realize the medium access, but to reduce the multi-path interference. The first DS realizations offered data rates up to 1 or 2Mbps working in the 2.4GHz frequency band. Shortly after that, 802.11b was introduced offering data rates up to 11Mbps. Three fully orthogonal channels can be used to avoid interference with neighbors. As the 2.4GHz frequency band started to get congested, IEEE802.11a was introduced working in the 5GHz band. Now more orthogonal channels are available (depending on the region, up to 12 for indoor use) and data rates up to 54Mbps are supported. Besides the change of the frequency band from 2.4 to 5GHz, 802.11a is using OFDM for higher spectral efficiency. As the OFDM technology has proven benefits over the DS technique, IEEE802.11g was introduced to using OFDM also in the 2.4GHz band. As both, 802.11b and 802.11g, work in the same frequency band and have the same MAC protocol, those two technologies are often implemented on the same chipset nowadays.

For both, 802.11a and 802.11g, the maximum data rate of 54Mbps will only be achieved if the communicating stations have a high signal–to–noise ratio (SNR) on their communication link. Loosely speaking, the SNR decreases with a larger distance between the stations. Other factors such as shadowing, multi–path, interference etc. are also playing a role, but to keep it simple we mostly refer to the distance. Depending on the prevailing SNR values the stations will adapt their modulation and coding scheme. Therefore, the data rate decreases with a lower SNR, which in turn depends on the distance between the stations.

Supporting the developments on the upcoming chapters, we focus next on the medium access control (MAC) of IEEE802.11 in the distributed coordination function (DCF). The MAC is based on carrier sense multiple access with collision avoidance (CSMA/CA). This means all participating stations sense the medium to understand whether it is already busy or not. If the medium is occupied the sensing station would not send at all to avoid collisions. Collisions occur if more than one station is using the wireless medium and the sender would receive multiple overlay signals that it cannot decode successfully. Whenever the medium is sensed as free, the station prepares to transmit on the medium. As there are possibly other stations also waiting to use the medium, each station has to wait for a certain time before transmitting anything. Those waiting times are different from station to station. The station with the smallest waiting time will send first. Now the medium is busy again and the other stations will freeze at this point in time waiting for the next free period to come. When a station has sent a packet it will wait for an acknowledgment from the counterpart communication device. In case there is no acknowledgment, the station will assume that the previous transmission has undergone a collision with at least another station. Such collisions are still possible as two or more station could have had the same random timer. In that case the waiting time for the next packet will be doubled to produce more time diversity. In contrast to Bluetooth the channel is not equally slotted. A station will occupy the medium as long as it takes to transmit the packet. This time depends on the length of the packet and of the supported data rate. In addition to the sending time, the time for the acknowledgment needs to be taken under consideration. Between the sending and the acknowledgment, there is a small time when the medium is not used. To avoid the possibility that other stations start to transmit in those pause intervals, 802.11 has introduced different timers. The station responsible for sending acknowledgment will access the medium right after the reception of the packet. Other stations will need to wait a longer time and when this timer expires the acknowledgment is already on its way, stopping other stations accessing the medium.

As collisions reduce the efficiency of the communication system, in 802.11 RTS (ready to send) and CTS (clear to send) messages are used to avoid these potential collisions. RTS messages are sent out by the sending station to ask the receiver whether it is currently busy with other transmissions that the sending station is not aware of. When the receiving station is ready it will send the CTS message. After the successful reception of the CTS, the sending station starts to convey its message. By the RTS and CTS messages the neighboring stations are also informed that the medium will be busy for some time. At least with those stations that have either received the RTS or CTS message, no collisions should occur.

In IEEE802.11 unicast and broadcast messages can be used. Unicast is the communication between two stations (e.g., point-to–point), while broadcast describes the communication originated by one station and received by multiple stations (e.g., point–to–multipoint). The unicast data rate is determined by the SNR between the communication partners. In broadcast the data rates should be set according to the link with the weakest signal. Most 802.11 implementations use the lowest possible data rate whenever broadcast messages are used, while others use the highest data rate. Only a few chipsets allow the data rate to be set in the case of broadcast. A combination of unicast and multicast transmission is the opportunistic listening approach. Here two stations are communicating in the normal unicast mode and the neighboring devices are overhearing the communication. This approach has some advantages over the broadcast as at least one acknowledgment will be received by the sender. The interested reader is referred to [2].

4.4 Combined Air Interface

So far in this chapter we assumed that cellular and short–range communications operate in different frequency bands not effecting each other. As this is desirable from the interference point of view, the drawback is that there are two air interfaces draining the battery. As we can see in Chapter 9 there is still energy that can be saved by the mobile cloud concept even if two air interfaces are powered on. But in the future the mobile cloud can be realized by one air interface. This approach will be even more energy efficient. In [3–6] we have already presented some ideas towards a combined air interface in the context of cognitive radio and cognitive networks. The main idea was to use the flexibility of the OFDM air interface and assign dynamically OFDM subcarriers to the cellular and the short–range communication in dependency of the service and the number of cooperating devices in the mobile cloud. In Figure 4.2 the comparison of a standalone device (on the top) with 24 subcarriers is compared with the case of two devices receiving partial information from the overlay network and exchanging it via the short–range communication link. The subcarriers in the example (bottom) show the subcarrier usage for the overlay (12) as well as sending (2) and receiving (2) on the short–range link (for one device). The example shows that now eight subcarriers are unused and contribute to the reduction in complexity and energy consumption.

The complexity of an OFDM air interface is given by $N \cdot log(N)$ with N as the number of subcarriers used. Assuming that a mobile device needs to receive N subcarriers to receive a certain service, e.g., video multicast, the question is how many subcarriers would be used for the mobile cloud concept. The subcarriers needed for the mobile cloud depends on the number of devices J in the mobile cloud and the speed gain Z between short–range and cellular. So Z describes how much faster is the short–range link to the cellular link. As the distance between two devices in the Device–to–Device (D2D) context is lower than the distance of the cellular link, higher data rates will be achieved on the direct link compared with the cellular link. The number of subcarriers N_{MC} that will be used in the mobile cloud is then given by

$$N_{MC} = N \cdot \left(\frac{1}{J} + \frac{1}{Z}\right). \tag{4.1}$$

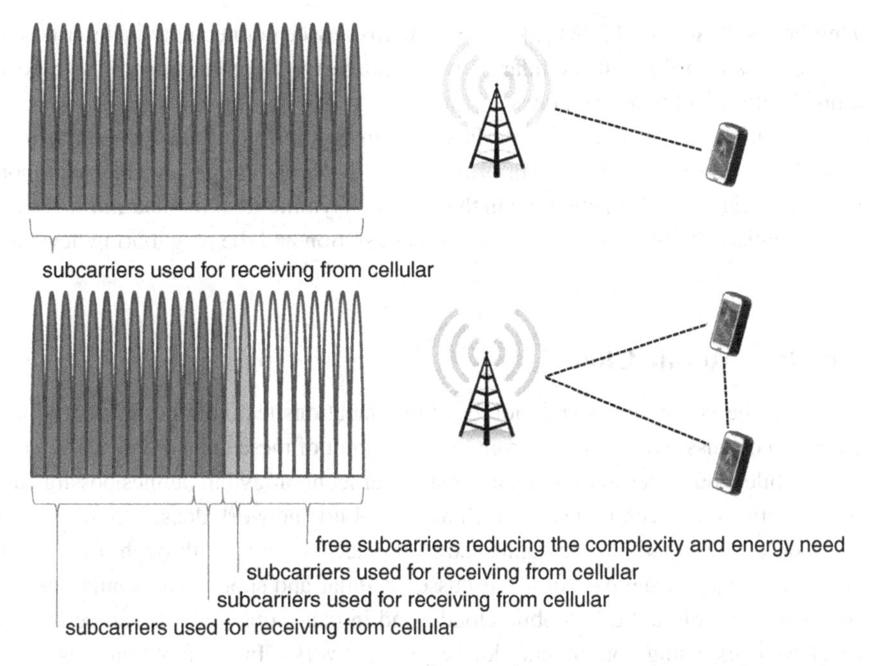

Figure 4.2 Example of dynamic OFDM subcarrier assignment compared with standalone device.

Compared with N subcarriers a standalone device would need, a device in the mobile cloud would receive only a portion of those subcarriers from the overlay network. And the number of subcarriers that are used to receive from the overlay depends on the number of cooperating devices in the cloud and equals N/J. This received portion will be send out to the other devices in the mobile cloud and, as the short–range is Z times faster than the cellular, $N/(JZ)$ subcarriers are used for that. As not the full information is received from the overlay, the device needs to retrieve the missing pieces from the other $J-1$ devices in the cloud. For this the device will use $N \cdot (J-1)/(JZ)$ subcarriers. Adding these three parts up will result in Equation 4.1.

From the network operator's perspective the number of subcarriers that will be given out to support mobile clouds is potentially larger than N but they will be used for a smaller amount of time. If the mobile cloud concept is not used, the N subcarriers will carry the service with additional redundancy to cover for the potential losses on the wireless medium. This redundancy is increasing with an increasing number of users. If the mobile cloud concept is applied the additional redundancy is less as fewer users are receiving the full information. At this point we do not give a full performance evaluation for the problem here (as it depends on the number of clouds that have been formatted with different numbers of users in the actual mobile cloud).

But also in LTE–A Device–to–Device communication in the LTE band is envisioned. Here the network operator acts as a spectrum manager applying resources for a direct communication among devices in the cell and service discovery from the network side is discussed. In 3GPP

Rel–10 the Local IP Access (LIPA) [7] the Device–to–Device connectivity is described. The main motivation is to offload the cellular network and perform communication with devices in close proximity with higher data rates.

In [8] an even more advanced air interface was introduced. The idea enabled a mobile device member to transmit different information to the base station and the neighboring mobile device members at the same time in the uplink. Asymmetrical modulation was used in order to differentiate the information flow to the base station and the neighboring devices.

4.5 Building Mobile Clouds

After reviewing the key wireless and mobile communications technologies currently in use, this section will discuss mobile clouds from the standpoint of these technologies. The generic concept of mobile cloud does not rely on any particular technology but defines possible direct wireless interactions between nodes as much as any ad hoc network does, and also assumes that each node of the cloud can be connected to the access network through a base station or access point. Certainly the previously discussed cellular and short–range communications technologies are suitable to build mobile clouds, and in fact mobile clouds can be built with current technologies, using commercial devices and networks. In principle there is no need to develop new equipment particularly suited to mobile clouds, though future developments could be better optimized for mobile cloud operations. Bluetooth and its topology limitations is an example of current technology that can be used as a building block of mobile clouds but its use will result in a cloud with non–optimal performance and inefficient utilization of the cloud and network resources. Cellular and short–range networks appear to be the natural component network technologies of mobile clouds. However, this is an assertion that holds today, but in the future there will be different options to build mobile clouds. Figure 4.3 depicts two basic approaches to build a mobile cloud, namely based on multiple network (and hence air interface) technologies (upper picture) as well as based on single network technology (lower picture).

Current feature and smart phones have integrated multiple air interfaces, at the best with several interfaces for short–range, and several for cellular access. Short–range and cellular interfaces can be considered to be orthogonal to each other, as they do not cause mutual interference. The air interfaces are typically implemented on different chip sets. These interfaces are currently not used jointly, but the most appropriate one is selected for a given task, or in a particular operating scenario. However, it is worth noticing that even with this current technology full fledged mobile clouds can be readily implemented. On the other hand, mobile clouds can also be implemented with a single air interface, based on a unique technology. OFDM lends itself to achieving this task easily, as different subcarriers can be utilized for local (short–range) connectivity and for centralized access to the overlay network. Clearly, orthogonality between cellular and short–range links are provided by the OFDM subcarriers, in this example. The number of subcarriers allocated for short–range and cellular access can be dynamically changed based on the prevailing needs. Current technology developments, notably LTE–A, supports this approach with their Device–to–Device approach allowing direct communication between devices using basically the same air interface used for cellular communications. The

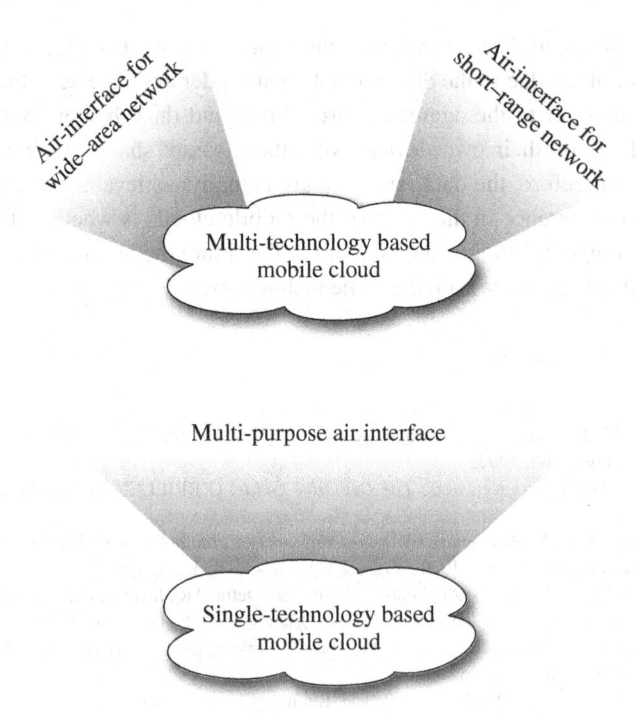

Figure 4.3 Implementing a mobile cloud: multi– and single–technology approaches.

single air interface can be also realized using software-defined radio technology. In addition to smaller cost and reduced real state requirements for its implementation, a single air interface has the advantage of consuming less energy than having multiple air interfaces active while the mobile cloud is operating.

In the future we can even expect that not all air interfaces on mobile devices are based on radio technology. Visible light communications (VLC) is a potentially strong candidate to complement radio systems in an important number of scenarios, from everyday typical scenarios to interference–sensitive environments. VLC is based on the use of white light emitting diodes (LEDs), which are primarily installed to provide illumination but at the same time can be used to provide downlink connectivity by data–modulating the white light. In addition to providing a radio–free communication, VLC is a low–power consumption and low–cost solution that also provides secure communications within the room the system is installed in. In the context of mobile clouds VLC suits as the access network that provides connectivity to the cloud.

4.6 Conclusion

In this chapter different technologies for cellular and short–range communications have been introduced. The evolution of data rate in cellular systems is a steady process offering up to

1Gbps with LTE–advanced. Data rates in the short–range are typically ten times higher than in the cellular technology due to the distance between sender and receiver. But the maximum supported data rates refer to the aggregate throughput, and thus they are not the data rates a single user will see on their own device. All data rates are shared among other users in the coverage cell. Therefore, the data rates that are actually achievable depend on the given scenario. The main difference in the setup of the mobile clouds is whether the spectrum for cellular and short–range is orthogonal or shared. In the future the air interfaces will be able to support cellular and short–range with the same technology.

References

[1] IEEE. Wireless LAN Medium Access Control (MAC) and Physical Layer (PHY) Specification. Technical Report 802.11, IEEE, Piscataway, NJ, 1997.

[2] Matthew Gast. *802.11 Wireless Networks: The Definitive Guide*. O'REILLEY, 2nd edition, May 2005. ISBN 978-0-596-10052-0.

[3] J.M. Kristensen and F.H.P. Fitzek. *Cognitive Wireless Networks – Cellular Controlled P2P Communication Using Software Defined Radio*, ISBN 978-1-4020-5978-0 22, pages 435–455. Springer, 2007.

[4] J.M. Kristensen and F.H.P. Fitzek. The Application of Software Defined Radio in a Cooperative Wireless Network. In *2006 Software Defined Radio Technical Conference*, Orlando, Florida, USA, November 2006. SDR Forum.

[5] J.M. Kristensen and F.H.P. Fitzek. Cooperative Wireless Networking Using Software Defined Radio. In *International OFDM Workshop*, August 2006.

[6] J.M. Kristensen, F.H.P. Fitzek, P. Koch and R. Prasad. Reducing Computational Complexity in Software Defined Radio Using Cooperative Wireless Networks. In *International Symposium on Wireless Personal Multimedia Communications (WPMC'05)*, Aalborg, Denmark, September 2005.

[7] 3GPP Technical Specification Group Services and System Aspects. Local IP Access and Selected IP Traffic Offload (LIPA-SIPTO). Technical Report 3GPP TR 23.829 V10.0.1 (2011-10), 3rd Generation Partnership Project, 2011.

[8] Q. Zhang and F.H.P. Fitzek. Asymmetrical Modulation for Uplink Communication in Cooperative Networks. In *IEEE International Conference on Communications (ICC 2008) – CoCoNet Workshop*, May 2008.

5

Network Coding for Mobile Clouds

Creativity is the ability to introduce order into the randomness of nature.

Eric Hoffer

This chapter introduces the main concept of network coding and underlines the importance of this disruptive key technology for mobile clouds. With reference to the previous chapters, the advantage of network coding is shown for efficient data dissemination in meshed networks, data storage and security. The aim of this chapter is to make the reader familiar with network coding concepts and principles without going through the details of the full theory behind it. This approach is enough to show the reader the excellent match between mobile clouds' architecture and associated information flow on the one hand, and network coding efficiency across these distributed nodes on the other hand.

5.1 Introduction to Network Coding

We have emphasized the importance of network coding for mobile clouds in previous chapters and will rely on the capabilities of network coding in the following chapters as well. In this chapter, we formalize the basic concepts of network coding. We approach network coding in an intuitive manner, aiming at motivating users to understand its underlying principles and to learn how network coding can be applied to mobile clouds. Network coding can be used for many purposes such as security, file distribution and distributed storage, but this chapter focuses on the data throughput improvements and reduction in control information for the mobile cloud communication. These improvements will also have a significant impact on

Mobile Clouds: Exploiting Distributed Resources in Wireless, Mobile and Social Networks, First Edition.
Frank H.P. Fitzek and Marcos D. Katz.

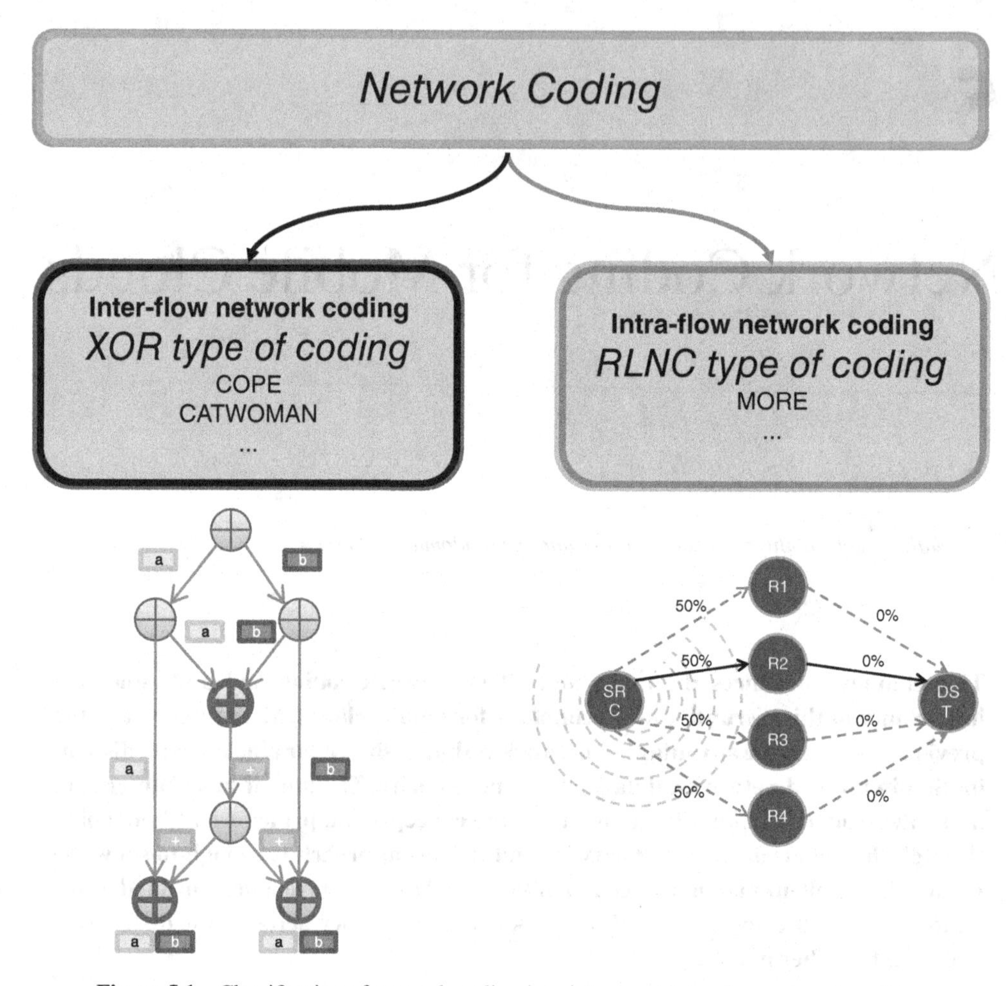

Figure 5.1 Classification of network coding into inter– and intra–flow network coding.

energy savings. The two main concepts of network coding are inter– and intra–flow network coding. This network coding classification together with the representative initiatives in each area are depicted in Figure 5.1. We will start to explain inter–flow network coding as this is the most widespread concept for applications right now. But we note that intra–flow coding has more potential and perfectly well suits the characteristics of mobile clouds.

5.2 Inter–Flow Network Coding

First we introduce network coding in the same way it was introduced to the research community by Ahlswede [1] in 2000 who referred to it as inter–flow network coding. After the publication of that seminal work, network coding is conventionally introduced by using the well–known butterfly example shown in Figure 5.2. The butterfly example refers to a special network

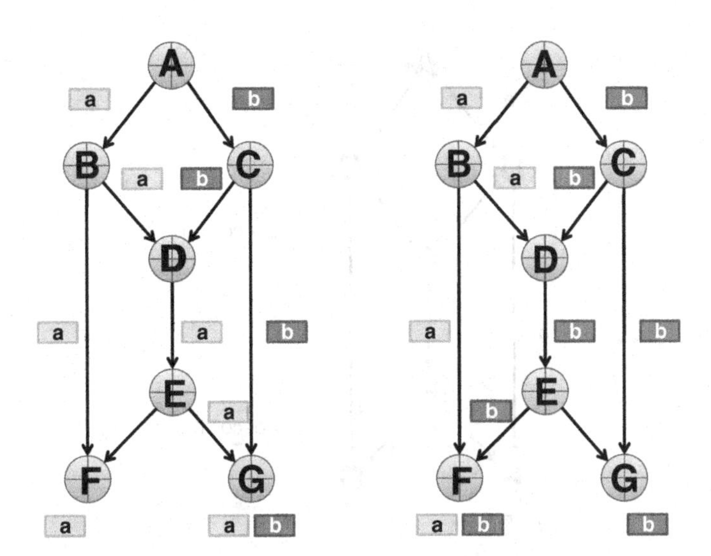

Figure 5.2 Butterfly network without network coding. The role of node D is crucial as its decision on which packet to forward (a or b) will determine which packets reach the destination nodes. Both approaches (left with packet a and right with packet b) are sub optimal.

topology, which helps to illustrate and understand the basic concept of network coding. At this point we would like to make clear that network coding is not limited to this kind of topology, but can be used in any arbitrary network topology. Even if this observation might be obvious to most of the readers, there are research works that identify butterfly sub–graphs in a larger arbitrary network in order to apply the findings of Ahlswede in each of the found butterflies. Even though the main breakthrough of network coding came with the work presented in [1] by Ahlswede, similar ideas had already been floating around for a long time [2] in the storage research community. The main contribution of [1] is therefore the proof that the min–cut max–flow capacity [3] can always be achieved in any arbitrary network topology for multicast transmissions.

Let us now explore the butterfly example in more detail. The butterfly scenario includes seven nodes that are connected as shown in Figure 5.2. The aim of source node A is to send two packets a and b of the same size to both destination nodes F and G over the given network topology. Let's assume that each link of that topology has the same capacity and is able to convey either packet a or b at a given time slot. Packets a and b are conveyed to node B and C, respectively. Node B will now forward the packet a to node D and F. At the same time node C will forward packet b to node D and G. After these transmissions the destination nodes F and G have already received one out of two packets. As can be easily seen in Figure 5.2 (left side) node D becomes the bottleneck of this topology as it receives both packets (a and b) but is only able to forward either a or b due to the channel capacity. Based on the decision of node D one of the destination nodes will get both packets and the other one will only get one packet.

In Figure 5.2 (left side), node D will forward packet a, so that at the end node G will have two packets, while node F will receive only one packet. In case of Figure 5.2 (right side),

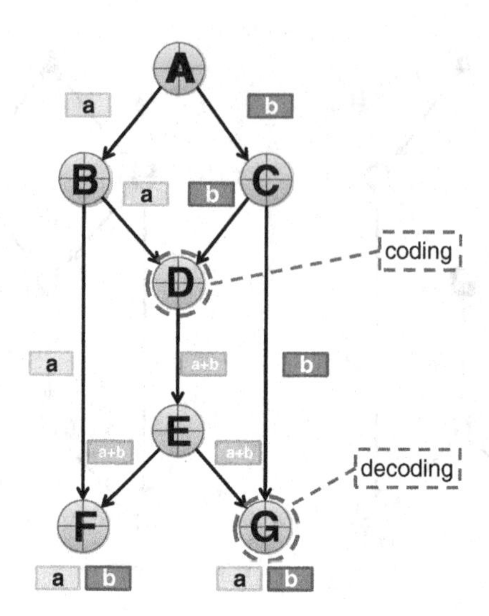

Figure 5.3 Butterfly network with network coding. Now node D computes a combined version of packet a and packet b.

node G will receive only one packet, while node F will receive two packets as node D has forwarded packet b instead of packet a. Whatever node D decides, one node will not receive the full information while the other node does. The throughput of the given topology is 1.5. This is calculated by summing up the throughput per node and dividing it by the number of destination nodes. This throughput is lower than the min–cut max–flow capacity [3] which would be two in this case. In a nutshell, the min–cut max flow states that each node would have a capacity of two if the other destination would not be present, assuming perfect routing and routing decisions. As we will see in the next paragraph, network coding will achieve the min–cut max flow capacity for the given example.

In Figure 5.3 the same network topology is given but now the nodes are able to perform network coding. Even though in principle each node is capable to perform network coding, we note already that not all nodes will necessarily perform network coding. Therefore, a network with both network coding–enabled and network coding–agnostic nodes is possible. Coming back to the example in Figure 5.3, the most interesting node is again node D. As before, it receives two packets and the outgoing capacity is large enough to forward only one packet. But this time, network coding kicks in. Node D produces a coded version of packet a and b. For illustration purposes we imagine that packets a and b are painted with yellow and blue colors, respectively. A coded version of both packets could be represented by a green color packet, referring to the mixing of blue and yellow. It is important to note that the size of the coded packet has exactly the same size of packet a or b and is therefore not just a concatenated version of both incoming packets. A green or coded packet, having the same size of either a or b, would be forwarded to node E which then also forwards that coded packet to both destination nodes F and G. Each destination node will receive two packets, namely one original packet (either a or b) and the coded packet. The coded packet alone is useless, but in combination

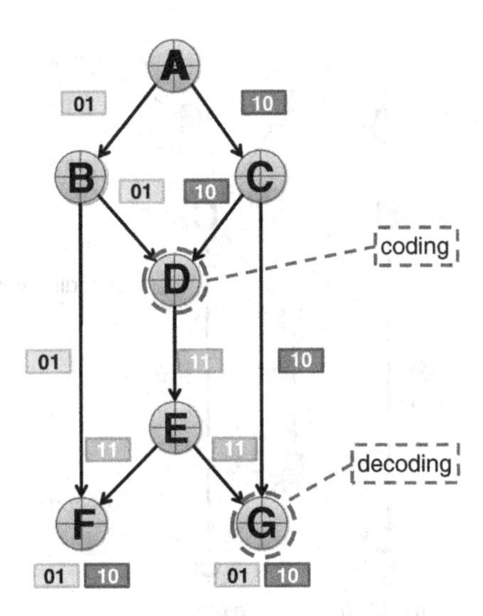

Figure 5.4 XOR (the Inequality function) operation in the butterfly.

with one of the original packets decoding can be carried out. In the case of node F, packet a is received from node B and the coded version comes from node E. To retrieve the original packet, node F decodes the green packet. As it already has the yellow packet it assumes that only a blue packet would make the coded packet green. At the end of the process, and unlike in the examples of Figure 5.2, both destination nodes F and G have received packets a and b. With this simple example it becomes clear for the reader that the decoding node needs some information about the topology of the network in order to make those assumptions. In other words, node F needs to know that the packet coming from node E is a coded version and not just a forward packet. Furthermore, information on which packets are coded together is needed too. We come back to this problem later on. Following this illustrative example, a more technical description of this type of network coding is given in the following.

Coming back to node D, the binary representation of packet a and b is coded by a bit–wise XOR operation, as given in Figure 5.4. Thus, the coding would be a bit–wise operation of packet a (represented by 01) and packet b (represented by 10). The result of this coding operation results in a packet with the same size represented by 11. Destination nodes will later decode the received data. The decoding operation is again the bit–wise operation of packet a (represented by 01) with the coded packet (represented by 11) and results in packet b (represented by 10). This simple example illustrates the basics of the coding and decoding. Furthermore, it also shows that each packet in this example had only two bits, even the coded version. It comes without saying that this approach works for packets of any length, but we limit ourselves to two bits for simplicity. It is important to notice that we have to introduce some overhead to convey information to the destination on how coding has been performed, as the butterfly is only one possible topology. As we stated above, the knowledge on which packets are coded with each other needs to be added to each packet and is referred to as encoding vector. In cases where the topology never changes this overhead is not needed, but

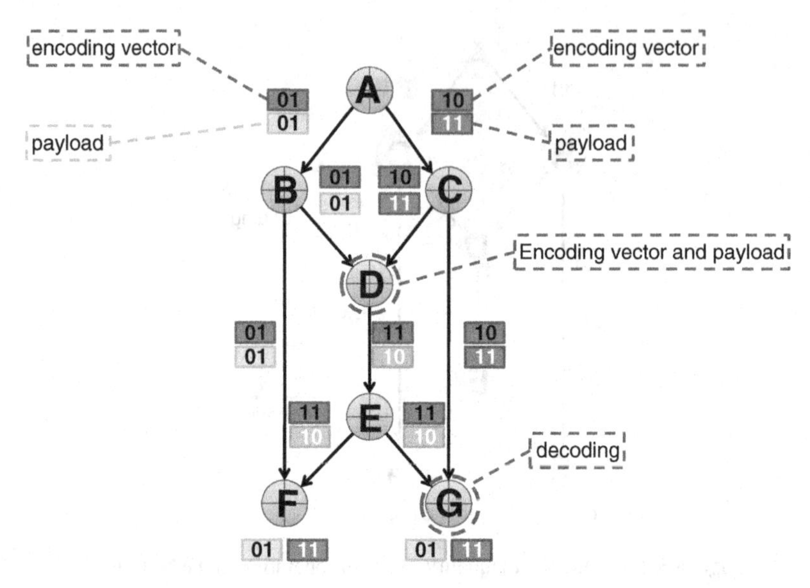

Figure 5.5 Butterfly example including the encoding vectors.

for topologies where the topology is changing, the information has to be carried with the packets. The encoding vector has to be transmitted with the packets and reduces the efficiency of the network coding approach. The amount of required overhead can be reduced, and some studies consider techniques to do so, but in most cases the impact of overhead is not highly relevant. E.g., if we consider the maximum transmission unit (MTU) of 1500 byte in the Ethernet and compare it with the two bits of the butterfly example, it is clearly negligible. Nevertheless, if we consider intra–flow network coding, as introduction in Section 5.4, the overhead will be larger most likely.

In Figure 5.5 the encoding vector is given for each packet. The encoding vector is used already at node *A*. Here the first packet *a* gets the identifier 01 and the second packet *b* the identifier 10. The encoding vector undergoes the same XOR procedure as the payload for both, coding and decoding. Therefore, the identifier for the coded packets is 11. Note that in case we would need to address a third packet, which we do not need here, the identifier would be 100, whereas 011 is a coded packet of 001 and 010. At this point we see that the overhead for the encoding vector is 50%. But this is just a simple illustrative example. In the case of the butterfly the overhead would always be the two bits, while the payload can have any size in bits. The larger the payload size, the smaller is the impact of the overhead. On the other hand, the use of encoding vectors allows the usage of network coding in any arbitrary network topology. Figure 5.5 shows the coding of the payload and the encoding vector for the butterfly example. We changed the payload for packet *b* (from 10 to 11 in contrast to the first example) in order to make this example more valuable. Inter–flow network coding is not just a theoretical concept but has already been implemented in wireless meshed networks. Here we refer to two projects called COPE [4] and CATWOMAN [5]. As shown in Figure 5.1, COPE and CATWOMAN are the two most prominent implementations of inter–flow network coding on commercial platforms and they will be discussed in more detail later in this chapter.

5.3 Inter–Flow Network Coding for User Cooperation in Mobile Clouds

In this section, the interplay between inter–flow network coding and mobile clouds is described. In Figure 5.6 we give a very simplistic and practical example on how both approaches can be combined. We assume an overlay network, in this case realized by LTE, introducing two packets a and b into a mobile cloud. The mobile cloud is composed of three mobile devices that are partially connected using the Wi–Fi technology. In the given scenario, all three nodes would like to receive both packets. In the real world, this could correspond to a case where all members of the mobile cloud are interested in receiving the same content. In a conventional manner, the overlay network could send both packets directly to the mobile users, but this would lead to increased bandwidth and energy usage (see Chapter 9). Therefore, we will exploit the direct connection among the mobile users to reduce the load of the overlay network. As given in Figure 5.6, the middle node is connected to the other two nodes, while the outer nodes need the middle node to communicate with each other and have no chance to hear each other. This topology is often referred to as Alice and Bob scenario, where Alice and Bob will communicate via a relay with each other. We assume that the outer nodes, namely Alice and Bob, are receiving the packets from the overlay network and the middle node is undergoing the relaying task without receiving anything from the overlay network. Therefore, this approach is referred to as pure relaying. Pure relaying requires a total of four transmissions in order to exchange two packets between Alice and Bob, as shown in Figure 5.6. The outer nodes

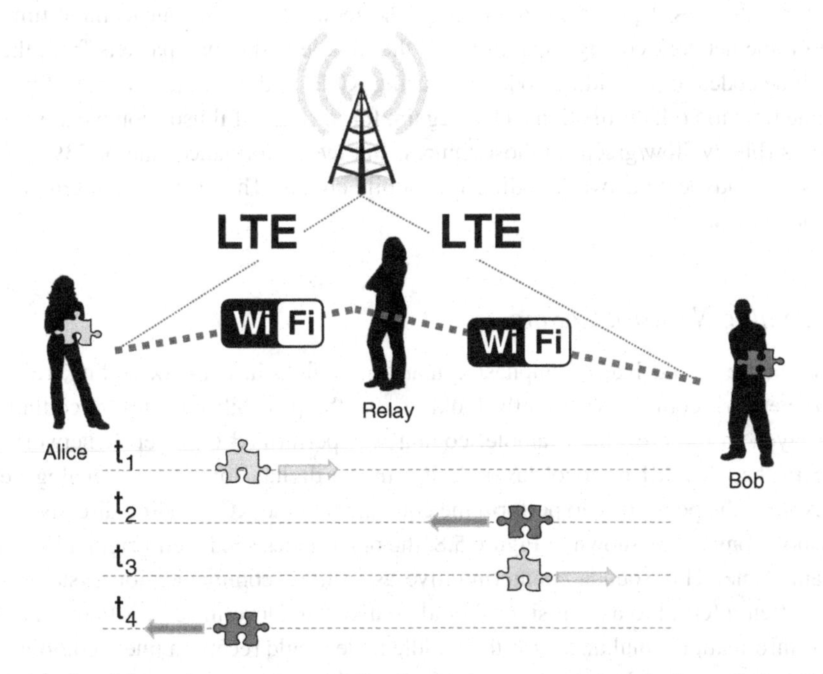

Figure 5.6 Distribution of packets in a mobile cloud using conventional relaying techniques (without network coding).

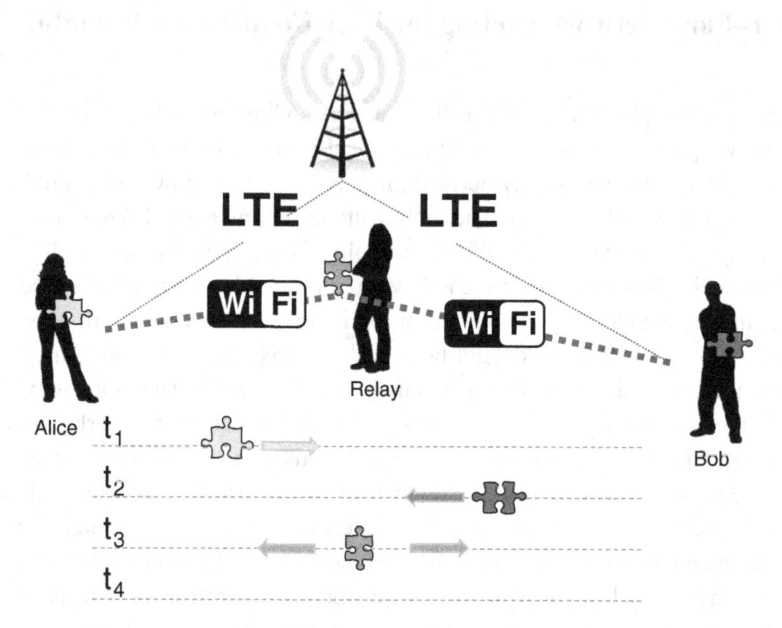

Figure 5.7 Distribution of packets in a mobile cloud using relaying techniques with network coding.

send their packets to the middle node, which is relaying two packets to the final destination. Using network coding as previously discussed, the number of overall transmissions is reduced to three time slots, as depicted in Figure 5.7. The reduction, from four to three time slots, is based on the network coding capabilities of the middle node. Two packets from the outer nodes will be coded at the middle node (e.g., using XOR) and the coded version is broadcast at the same time to both destinations. Once again, for the ease of illustration we are using the three colors (blue/yellow/green) in those figures. But the performance gain of 33% is not the main reason to advocate network coding for mobile clouds. The example chosen is just for illustration purposes.

5.3.1 *Analog Network Coding*

At this point we would like to emphasize that a new field in network coding, referred to as analog network coding, is currently looking into the possibility to do the coding at the physical layer. In the previous example, coding was performed on layer 3 (network layer) referring to the ISO/OSI protocol layer design in the digital domain. But analog network coding exploits the possibility to perform the equivalent to the XOR coding directly in the air in the analog domain. As shown in Figure 5.8, the outer nodes send their original information at the same time. This sounds counterintuitive as in most communication systems such a situation, often referred to as collision, should be avoided. Nevertheless, if both outer nodes send their information simultaneously, the middle node would receive a linear combination of both packets in the signal domain. This combination will then be broadcast to both destination

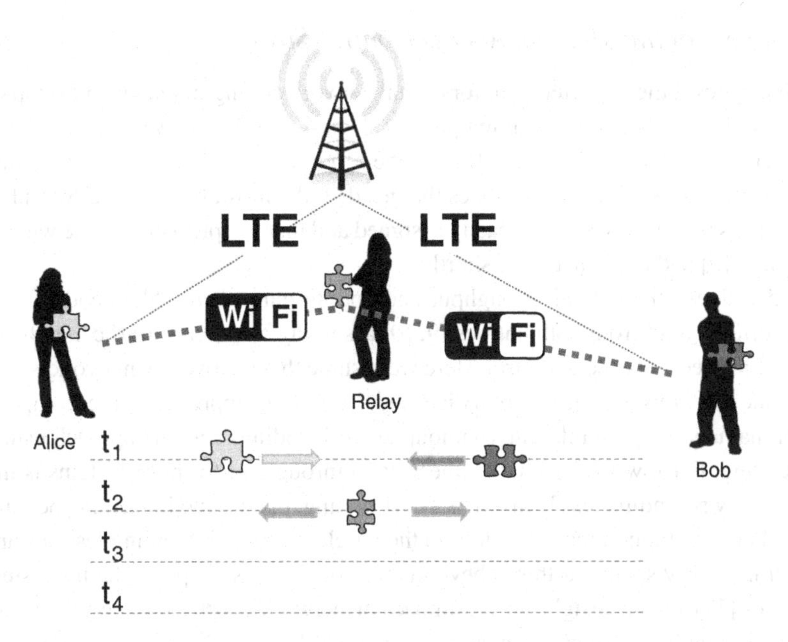

Figure 5.8 Distribution of packets in a mobile cloud using relaying techniques with analog network coding.

nodes. Analog network coding would reduce the amount of transmissions further down to two. So this approach would not differ from a scenario where the two outer nodes could have been communicating with each other directly using just two time slots. While the packets are combined in the digital domain, signals are coded in the analog domain.

5.3.2 Comparison of Analog and Digital Network Coding

Let's now compare these three approaches. Both network coding cases yield better results in terms of used time slots for packet transmissions compared with the conventional approach which is based only on relaying. While digital network coding adds only little complexity, analog network coding is currently harder to implement and puts some conditions on the prevailing signal–to–noise–ratio. Furthermore, it has to be noted that the role of the middle node is different for both coding schemes. With digital network coding the middle node is a full member of the mobile cloud as it receives all information. In the case of analog network coding, the middle node is degraded to a simple relay, for which the node is not able to use the relayed information for itself. The latter one is beneficial with respect to security and privacy, if the outer nodes would not like the middle node to understand the undergoing communication. But for the operation of the mobile cloud, it is of utmost importance to show each member a clear benefit in participation (see Chapter 8). Currently the research community is looking into layer 3 network coding as well as analog network coding. But most implementation efforts are made in the digital domain coding [4, 5] rather than in the analog domain [6–8].

5.3.3 *Impact of the Medium Access Control Strategy*

As explained above, the expected gain for digital network coding in the given example is 33% as we reduce the number of transmitted packets from 4 to 3. But if we consider the medium access scheme then the results will differ. In the aforementioned example, we assumed that the wireless medium is given to the nodes that needs it the most, but in IEEE 802.11 medium access all nodes will get the same capacity assigned and that is a problem for the whole system as is shown in [5] and explained here shortly.

Figure 5.9 shows theoretical throughput and coding gain of the Alice and Bob scenario versus the offered load from both entities [9, 10] assuming symmetric traffic and the usage of IEEE802.11 as medium access control. Here we assume that the overlay network is constantly filling up the transmission queue of Alice and Bob. Two approaches are compared with each other, namely the pure relaying (without network coding) and an approach using digital network coding. For low load scenarios the system throughput for both systems is the same. Even if the relay is sending less by using network coding, that has no impact on the throughput as long as there are enough resources left on the wireless channel. Sending less has an impact on the overall energy spent but this is beyond the scope of this chapter. The interested reader is referred to [11] for an insight on the impact of network coding on energy consumption. With an increasing load, the performance of the pure relaying scheme degrades while that of network coding still increases. At a certain load, both approaches stabilize their performance. Interestingly the coding gain exceeds the 33% we discussed beforehand and reaches values up

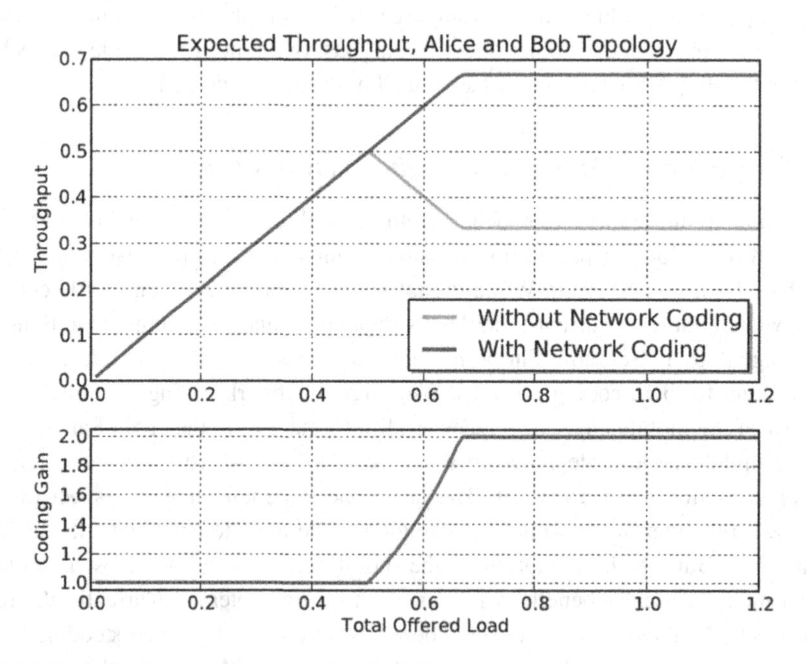

Figure 5.9 Throughput and coding gain for the Alice and Bob scenario with and without network coding [9, 10].

to 100%, so doubling the system throughput for this simple example. Let's first consider the fact that, in the working range described below, both approaches yield the same performance (until Alice and Bob are introducing a load of 50% of the nominal wireless channel capacity). Even if the throughput is the same for that given traffic load limit, from the energy perspective the relay node has to transmit twice as much and this might cause problems if the relay node would be battery driven. Increasing the offered load further shows impact in the system throughput. The reason behind this is the way the MAC of IEEE802.11 works providing a fair share between all nodes. IEEE802.11 has neither information nor interest on what the relay node is doing for Alice and Bob and therefore is not assigning more wireless spectrum to the relay. Loosely speaking, Alice and Bob are shooting themselves in the foot by *stealing* the capacity from the middle node. In an ideal case where a MAC would have been optimized for the relaying node, the coding gain would not exceed the 33% for the given scenario. But currently the WLAN world is dominated by IEEE802.11 technology.

The theoretical results shown in Figure 5.9 are well in line with the implementation efforts in the research community. XOR coding was applied to wireless meshed networks on MIT campus in [4] by Katti et al. introducing the COPE mechanism. COPE is not only the XOR type of packet coding it also deals with implementation towards the IEEE802.11 standard. The main outcome of the work was a 3–4-fold increase in capacity for the wireless meshed network using network coding compared with a non–coded system under high load scenarios and UDP traffic. In [12], the authors applied a similar COPE mechanism to the Nokia N810 platform for efficient data dissemination. In [13] a video shows how this approach works and gives some insights about the potential gain. In [5] the CATWOMAN approach is introduced. CATWOMAN combines the XOR coding with the BATMAN [14] routing protocol. CATWOMAN is implemented on commercial Wi–Fi hotspots measuring the performance of network coding for simple setups such as Alice and Bob, the cross topologies and many more. CATWOMAN could show that the real performance gain for Alice and Bob is 64%. Therefore, it proved that the performance is higher than the 25% but less than the 100% from the theoretical work. The main reason not reaching the 100% is the asymmetry of the traffic coming from Alice and Bob, the medium access control itself and the characteristics of the wireless channels. Further implementation show the gain of more advanced topologies such as the chain topology, the X–topology, and the cross topology [15, 16].

The simple form of network coding using XOR has some drawbacks. While the simplicity allows quick implementation, the nodes performing network coding need to understand which packets need to be coded in order to be efficient. If this is achieved by overwhelming signaling, the performance will be degraded. Owing to the mobility of the users in a mobile cloud, the topology is prone to changes in channel quality or route selection all the time. As given in Figure 5.10 four users are fully connected (left side) and in case of packet erasures retransmissions will occur without the use of network coding each user is responsible to do the number of necessary retransmissions. On the right-hand side of Figure 5.10, the topology has changed and the centered user has a very good connection to the outer nodes. Again, each user could use just retransmission schemes but network coding could help. Here network coding would be used by the centered node to retransmit a coded version of packets coming from the outer nodes.

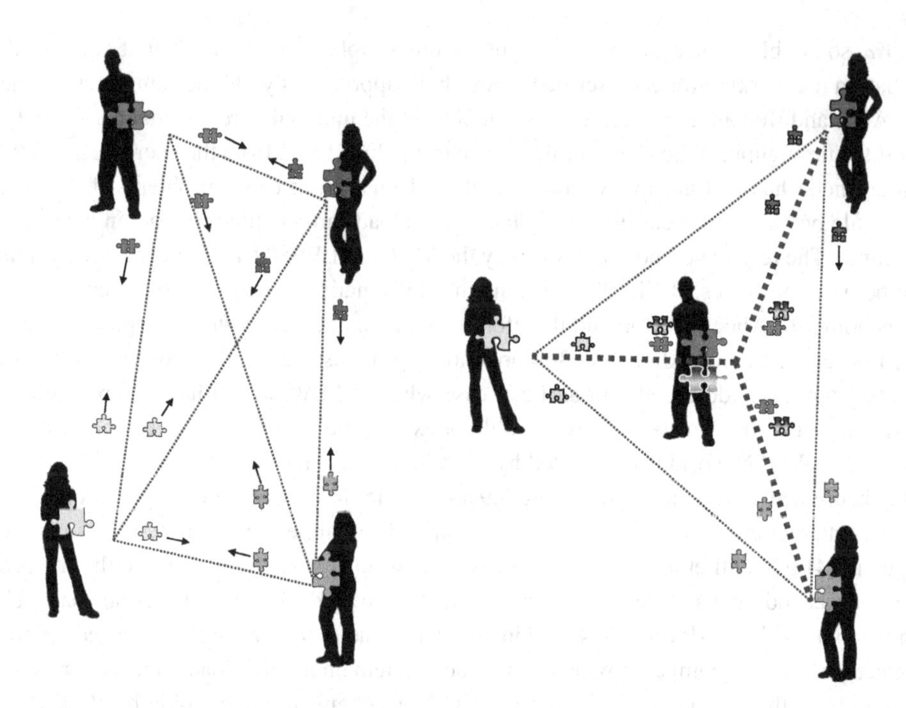

Figure 5.10 More advanced mobile cloud topologies and the need for network coding.

5.4 Intra–Flow Network Coding

A huge milestone in the short history of network coding was introduced by Ho et al. [17] with the concept of random linear network coding (RLNC), which is the base technology for intra–flow network coding. Instead of designing carefully the coding vectors as required by inter–flow network coding, the authors of [17] have shown that the coding coefficients can be picked randomly without decreasing the performance of the system significantly, referring to this technique as random linear network coding (RLNC). Inter–flow network coding with complete global knowledge over all nodes will slightly outperform its intra–flow counterpart in most cases, but the price to control a full system and disseminate the coding parameters in the former approach can be prohibitively high, especially if the network becomes larger.

Figure 5.11 shows the principle of RLNC. The figure shows the example of two relay nodes (R) that receive broadcast information from the source node (S) and are willing to forward that information to the destination node (D). We consider only two packets a and b and investigate the probability of sending those two packets in two time slots. If both packets are received successfully from the source, the relays should forward those packets. The question arises as to which packets should be forwarded by which relay. In the presence of no coding, the relays might forward the same packet with probability 50%, which will be very inefficient. In order to avoid sending the same packet, the relays could overhear each other in order to transmit the orthogonal information: an assumption that does not always hold as the link between the relays

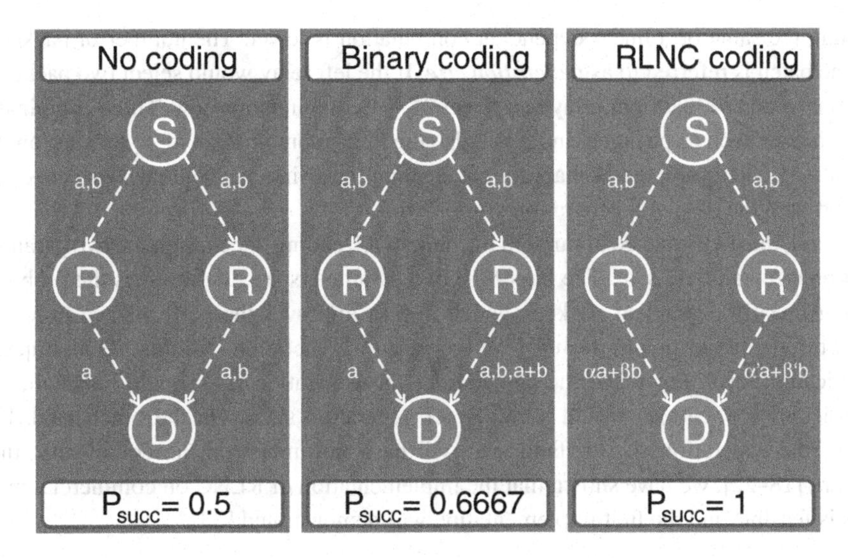

Figure 5.11 Network coding for a simple relaying example with two relays investigating the performance of network coding.

might be error–prone but more importantly packets will be queued in order to get assigned transmission slots (as for IEEE802.11) and no decision can be made after the queuing.

Using network coding (middle part of Figure 5.11), each relay could choose the encoding vector randomly. There are four possible encoding vectors, namely 00, 01, 10, and 11 using a small field size. The null vector 00 should be avoided as it does not contain any information. But this can be easily excluded. The encoding vectors 01 and 10 represent the original packet a and b, respectively. Besides the original packets a and b, there is one linear combination of both packets represented by the encoding vector 11. Both relays just choose randomly one packet that should be transmitted. That improves the probability to send both original packets in two time slots to 66%. But the probability can be increased to 100% if the coding coefficients are not limited to 0 and 1, but if we allow any number between 0 and 1. As both relays are choosing two random numbers for each packet the probability that they choose the same set becomes nearly zero.

But it is not just about choosing the same set but also *linear dependent* sets should be avoided. In practical implementations, the random numbers are mapped to finite field elements as floating point numbers are prone to rounding errors. Therefore, the probability of choosing the same set depends on the *field size*. Loosely speaking, the field size defines some granularity on how precisely the random numbers are generated. In our example in the middle, the finite field size was 2 (only 0 and 1 could be selected as coefficients), referred to as binary field. The binary field has some advantages when it comes to complexity, but the performance might be poor in some scenarios.

There is a second parameter that determines the performance of the system using intra–flow coding, referred to as *generation size*. If we come back to the binary case example, we had only two packets to send. If we would send more packets (e.g., a, b, c, and d) the probability of sending the same or a linear dependent combination is lower. The number of packets we set in one batch is referred to as *generation size*. If the left relay would select two packets out of the four (a and b), the right relay can select $2^4 - 1$ combinations now. Three combinations are bad choices such as packet a and b but also the linear combination of $a + b$ (just an XOR combination). Therefore, the probability of selecting linear independent packets is decreased compared with the case of a generation size of two.

In general, the performance of random linear network coding depends on the two parameters generation size and field size. The larger the two parameters are, the lower is the probability of having linearly dependent packets, which in turn improves the performance in terms of efficient throughput or goodput. But if we increase the generation size this has an impact on the application level. For real–time applications the generation size should be small, while delay–tolerant applications would allow larger generation sizes. On the other hand, larger values of the generation size or field size will have an impact on the complexity. In our prior work [18–22], we have shown that the implementation of RLNC on commercial mobile devices is feasible and the first implementation was demonstrated [23].

5.5 Intra–Flow Network Coding for User Cooperation in Mobile Clouds

In this section, we describe the three main fields where intra–flow network coding improves the performance of the mobile cloud, namely i.) exchange towards and in a mobile cloud, ii.) distributed storage in the mobile clouds, and finally iii.) security for the mobile cloud.

5.5.1 Exchange and Seeding Information for Mobile Clouds

The exchange of information within a mobile cloud as well as seeding data into the mobile cloud needs to be as efficient as possible in order to motivate as many users as possible to cooperate (see Chapter 8 for technology-enabled cooperation). With network coding, the number of exchanged packets can be reduced significantly and thereby increase the benefits of cooperation. In the following we describe how to seed information in the mobile cloud, and later we show how to exchange that information among the mobile devices.

Seeding Information in the Mobile Cloud

Figure 5.12 illustrates the use of network coding for seeding information into a mobile cloud. In the example, one base station transmits three packets to three devices. In order to reduce the number of transmissions, broadcast is used instead of unicast. The first packet that is transmitted in a broadcast fashion carries information x_1. Later packets two and three are also conveyed towards all users carrying x_2 and x_3, respectively. Owing to losses, some packets are not received by the users. In our example, we assume that each mobile device is losing a different packet. In case ARQ is the error recovery of choice, the originating base station would need to retransmit all three packets resulting in six overall transmissions, and a considerable

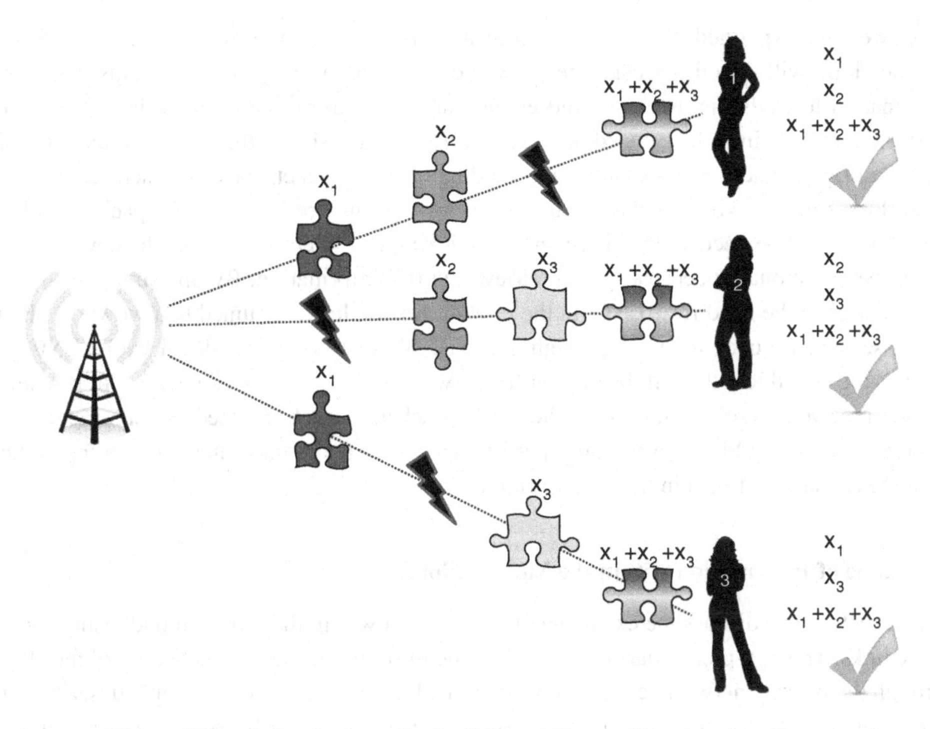

Figure 5.12 A simple example of reliable multicast with network coding.

waste of resources. The given example is the worst case scenario for one packet loss per device. The best case would be that all devices would lose the same packet, so that a single broadcast information might be sufficient. Nevertheless, with network coding only one additional packet is sent where all three packets are coded together. We note again that the coded packet has exactly the same length as the packets that have been previously sent. Coding is carried out here by simple modulo 2 operations (simple XOR as described before) over each bit of the three packets represented by $x_1 + x_2 + x_3$. The three packets have been received by all receivers, namely the two original packets and one coded packet. The coded packet enables each receiver to retrieve exactly that packet that has been missing. We have tacitly assumed that the base station knows about the losses and is doing the *right* coding decision. But if we extend the example to 16 packets that are broadcast in the beginning and the receivers would lose one packet again but the losses are not the same, then the base station would need to know about the exact losses to make the right coding decision. This illustrative example helps us to differentiate inter– and intra–flow network coding. In the example before, inter–flow network coding has been used. Using intra-flow network coding, the base station would code all packets together. The base station would then transmit as long as each receiver has enough packets, which depends on the scenario. If the receivers cannot cooperate each receiver needs as many linear independent packets as the generation size. If the receivers cooperate among each other, the group needs to ensure it has enough linear independent packets within the group.

As we have explained, the introduction of information from the overlay network into the mobile cloud will benefit by using network coding. If the overlay network wants to convey information towards the users it could either start transmitting the original data or it could start right away to introduce coded packets. In [20] we have shown that it is beneficial to start with the original packets and switch on the coding once all packets have been sent. In the case of no losses on the wireless channel the devices would just receive original packets and no decoding would be needed, thus reducing complexity. But in case of losses, the device needs to receive additional coded packets to retrieve the full information. By this simple example, we see already the need for RLNC. In the previous example we assumed that we know about the losses and we code together the right packets. Without going into detail with RLNC, all packets are coded together all the time in such a way that every received packet will be useful for the receiver. Here we note that other coding schemes such as Reed–Solomon codes or Fountain codes would achieve similar performance. The real gain of network coding comes with the exchange of data in the mobile cloud.

Exchange of Information within the Mobile Cloud

In this section, we discuss the exchange of information within the mobile cloud, using short–range links. This is typically carried out once some information has reached nodes of the cloud through the overlay network, e.g., through data seeding. The following example describes the benefit of network coding for user cooperation, focusing particularly on the exchange of data within the mobile cloud. Thus, we assume that the overlay network has already conveyed information to the users of the mobile cloud in such a way that the full information is available in some nodes of the cloud but not in each and every individual user yet. Thus, each user holds partial information and further exchange within the mobile cloud is needed before everyone has the full information. The exchange of data pieces can be done by using different cooperative strategies such as unicast, broadcast or network coding. The last one provides the best performance as we will discuss in the following.

In unicast communication mode, two peers in the mobile cloud will connect to each other and exchange non–redundant information. In case of erasures, retransmissions will occur. If two peers have exchanged their packets they have not necessarily obtained the full information, and must establish connections to new peers to obtain additional information. The peers continue this procedure until the full information is available.

In the broadcast communication mode, each node broadcasts its own information in addition to packets obtained from other nodes. In this mode a single transmission is received by multiple receivers compared with a single receiver in the unicast case. Owing to erasures in the broadcast case, retransmissions will occur. Repeating broadcast information differs in its usefulness among the receivers, some might already have the information while others see the conveyed information as novel. This problem is described in the literature as the coupon collector's problem.

In contrast, using network coding allows each user to recode its own packets before broadcasting it to the neighbors. Newly received information is used for recoding when available. In case of large field size, coded broadcast information will be seen as novel by each receiver as long as they still miss information. Therefore, the efficiency for the network coding–based approach will be higher. For the coding both inter– as well as intra–flow network coding can be used. Again the inter–flow network coding would need perfect knowledge about the missing packets of the neighboring nodes as implemented in [12]. For intra–flow network coding this knowledge is not needed. Here the only signalling message needed is one to show that the receiver has received sufficient information.

In order to understand the efficiency of the different schemes, we could summarize that each unicast transmission is only important for one receiver. In the case of broadcast, all receivers are potentially interested in a packet but redundant information reduces the efficiency. Only using network coding with an appropriate field size will guarantee that each transmission is relevant for each receiver that still suffers by losses, and redundant information will occur with a very low probability.

So far we assumed that all users in the mobile cloud will get the same amount of data for later exchange. But this might not be the optimal seeding strategy when the mobile cloud is established by a multi-hop network. The seeding of information into a mobile cloud becomes even harder if we have mobile nodes resulting in topologies with significant changes over time. In [24–28], we have used genetic algorithms together with network coding to find near optimal seeding strategies for cooperative download scenarios. The easiest solution would be to distribute the file over all nodes equally and rely on the later exchange among the nodes. But some nodes might be more suited to get a higher seeding than others. If we make an extensive search over all possibilities this will result in a large number of tests. Even worse the test needs to be repeated as soon as the topology changes. In the papers cited before, we could show that genetic algorithms achieve very good results with a low number of tests. Furthermore for smaller changes in the topology the number of additional tests is even lower than the number in the initial test phase.

The gain of using network coding over any other scheme will increase when the mobile cloud is not fully connected and some users can only reach each other through multi–hop links. The gain improves even more if the topology is changing over time due to wireless channel variations or node mobility. As random linear network coding does not require strict planning as the inter–flow network coding does, it can be used for highly dynamic topologies. Figure 5.13 depicts ten LEGO robots that are used to investigate high mobility meshed networks at Aalborg University. In [29] the results for content sharing among mobile devices with random linear network coding are given. The main finding of this study was that random linear network coding enabled the devices to share the content quicker than did any other scheme.

5.5.2 Distributed Storage in Mobile Clouds

Network coding can help to store information in the mobile cloud more efficiently in terms of reliability, used storage, and security (see Section 3.11 for the general idea). Let us assume one mobile device has important information to store. The file size is F (in bytes). There are different ways to store the data. First, the user could store the data in one place such as the

Figure 5.13 Network coding testbed for information exchange within a mobile cloud with high mobility [29].

Amazon cloud or locally on the phone. If we store it locally, the data is bound to the device and if the device is lost the data is lost too. If stored on the Amazon or similar cloud, this may involve a setup time and additional costs and therefore we look into storage on cooperative devices in the mobile cloud.

If cooperative users would agree to help each other to store information, one mobile phone could use the storage capacity of all cooperative mobile phones. As given in Figure 5.14, it is not important whether the mobile phones are connected via the short–range or any given overlay network. It is important that they are somehow connected and can exchange data. Figure 5.14 illustrates the seeding and retrieval phases. In the seeding phase the outer left device has 100% of the data and is seeding it into four different devices. As we apply random linear network coding for it, we could use any number of cooperative devices to full their storage in a meaningful way. In the retrieval phase, the outer right device has to connect three cooperative devices to restore the data. At this point, we just mention that the retrieving device could also get the data from all four devices if they are available and stop as soon as enough meaningful data has been received. One clear advantage is that leeching from four devices might be faster – an effect known from peer–to–peer downloading. The example in Figure 5.14 shows how the distributed storage might work, but in the following we derive some general formulations for it and compare it with other approaches.

Once there is the possibility to store data on distributed cooperative entities then the question is how we store the data such that we get it back with high reliability. As we rely on connected

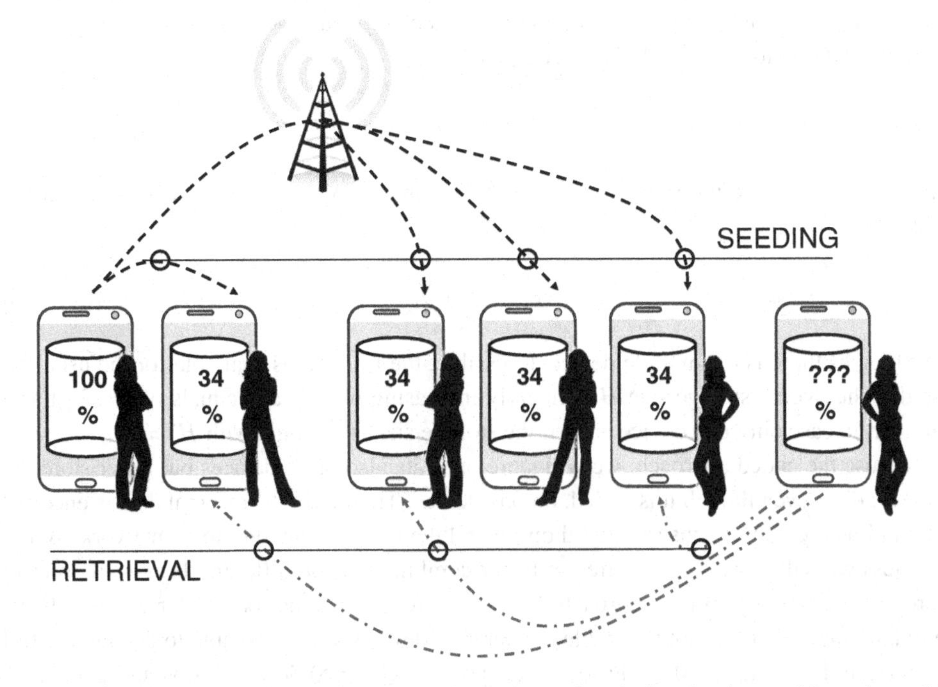

Figure 5.14 Example of distributed storage with seeding and retrieving data into and from the mobile cloud.

mobile devices, we cannot assume they are always available, due to missing connectivity or low battery status. Furthermore, the privacy of our data is important. Even though we are cooperative we would not like that others can read our data. Last but not least, the amount of storage that is used for the approach should be minimal, even though storage is cheap and available on the phones. In the following we explore three possible approaches:

One server approach In this approach we store the full data of size F on one cooperative device.

Duplicate server approach Here we store our data on multiple devices. Here the full amount of data F is stored on each device.

Coded approach The coded approach is based on the random linear network coding approach. Data will be coded and then distributed on several mobile phones. It is important to note that we will store less data on a cooperative device than the original data.

Those three approaches will be evaluated according to our metric focusing on reliability, storage, and security. The used equations are rather basic but give some insights about the usability of the approaches. Let us first evaluate the reliability A of the different approaches. For our calculation, we assume that we have N cooperative mobile devices and that the probability of availability of one cooperative device is p. In case there is only one device where the data is stored, the reliability A equals

$$A = 1 - p. \tag{5.1}$$

In order to improve the reliability we store the data on N multiple devices such that the reliability A becomes

$$A = 1 - p^N. \tag{5.2}$$

As long as there is one mobile device accessible, the full data is available for the user. We note that the overall storage size H is linearly increasing with N and equals $N \cdot F$. Just as a reminder, in case only one device is used the storage size is minimal with $H = F$.

If we use the coded approach, we will store our data also on N devices but never store the complete data, even though it is coded, on one device. Therefore, the original data is encoded first and linear combinations are stored on several devices. As random linear network coding is rateless we will always store different linear combinations on different devices in order to improve the efficiency of our approach. We also introduce another parameter T that reflects the maximum number of not reachable cooperative devices still being able to decode the full information. For instance, if we have five cooperative ($N = 5$) devices we would store 25% of the coded data on each device. Even if we cannot reach one device ($T = 1$), we still could get the full data decoded assuming the linear dependency is nonexistent, which is achieved by large field and generation sizes. Clearly, in the duplicate server approach $T = N - 1$.

For the coded approach, the reliability A becomes

$$A = 1 - \sum_{i=0}^{i=T} \binom{N}{i} \cdot p^{N-i} \cdot (1-p)^i. \tag{5.3}$$

The amount of data P (as a percentage of the original data size) that we have to store on each of the N cooperative devices is

$$P = 1/(N - T). \tag{5.4}$$

The overall storage size H that is needed for the distributed approach equals

$$H = N \cdot P \cdot F = N/(N - T) \cdot F. \tag{5.5}$$

If we come back to our example of five cooperative devices, the amount of data for each device is given in Table 5.1 as a function of T.

In Figure 5.15, the reliability A is given versus the number of cooperative mobile phones with a connectivity probability $p = 0.25$. For the one server approach, the reliability is always 75%. Duplicating the data to multiple devices achieves the highest reliability, achieving 98% reliability with three servers and 100% reliability with more servers. The coded approach with

Table 5.1 Overall storage usage for the distributed approach for five mobile phones.

N	T	P	Overall storage size H
5	0	20%	100%
5	1	25%	125%
5	2	33%	167%
5	3	50%	250%
5	4	100%	500%

$T = 1$ and $T = 3$ achieves less reliability than the duplicating approach and even less than the one server approach if not enough devices have been activated for distributed storage. But as soon as there are six ($T = 1$) or ten ($T = 3$) cooperative storage devices the distributed approach yields reliability values as the duplicated approach near to 100%.

The advantage of the distributed approach over the other approaches becomes evident if we look at the storage size needed and the confidentiality. First, we will have a look at the storage. We already mentioned that storage might be cheap and mobile devices have more than enough, but it is not only the amount of data that needs to be stored but also the amount that needs to be conveyed over the cooperative network, leading to an increased bandwidth usage and energy consumption. Obviously if we store our data on only one device we will use a certain storage space and other resources. We defined the resource usage RU to be equal to 1. If we duplicate the data to N devices the storage need becomes N–fold and therefore $RU = N$. This is clearly a huge disadvantage. The amount of data for the distributed approach is $RU = N/(N - T)$, for $N > T$. As given in Figure 5.16, the distributed approach is dramatically better than the

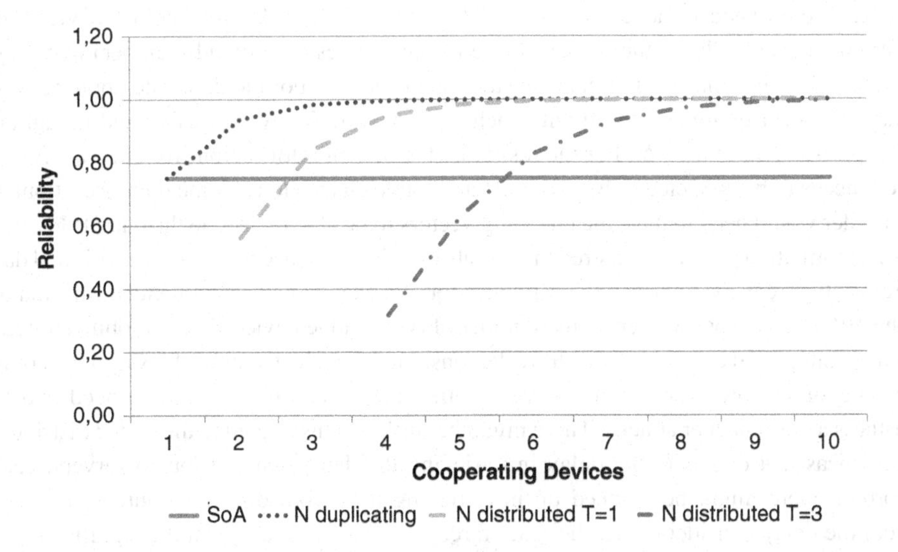

Figure 5.15 Reliability for the three different storage approaches with 25% outage.

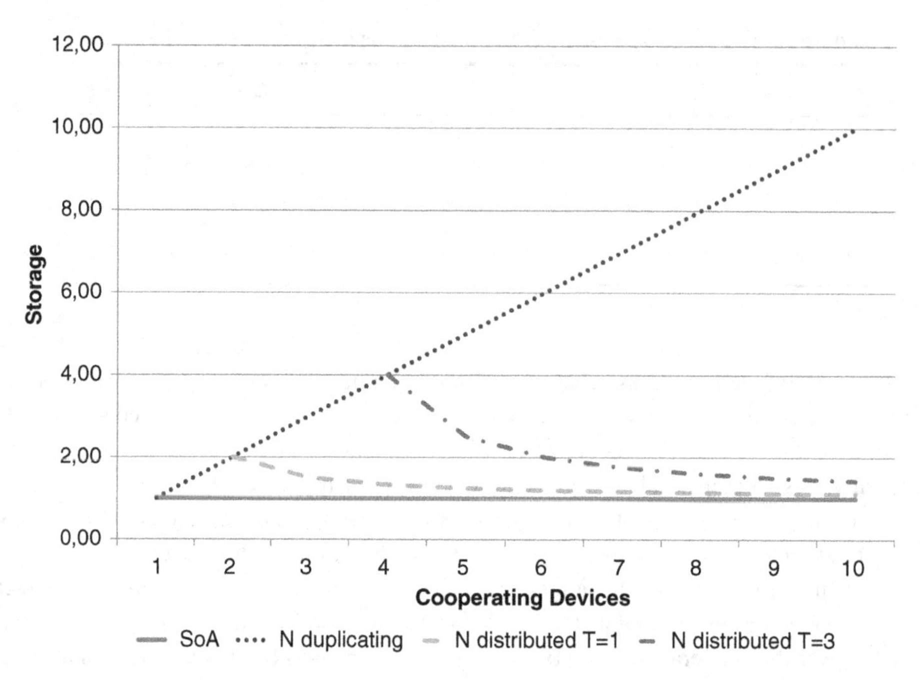

Figure 5.16 Allocated storage for the three different approaches.

duplication approach. For a large number of N the distributed approach becomes nearly as good as the one device approach in terms of resource usage.

Finally confidentiality is a very important issue whenever information is stored in a cloud. We make a very simple analysis based on the effort an undesired or non–target user (e.g., a hacker) would need to access the stored data. As we store data on mobile devices, the hacker could eventually be the owner of one of the devices (i.e., cloud member) that might be interested to see what kind of data we store on his device. For the duplicate approach both privacy and security are decreased with each additional device we are using and it degrades further proportionally to $1/N$. For the distributed approach information from at least $N - T$ devices needs to be collected to have at least a possible chance to hack the data. Additionally, the intruder would need to have the encoding vectors to be able to decode the data. If less than $N - T$ information parts are gathered there is absolutely no chance to recover the original data. Therefore Figure 5.17 describes the difficulty for a hacker to gather unintended information for the different approaches versus the number of cooperative devices in the mobile cloud.

In these simplified calculations we have demonstrated the effectiveness of distributed storage over state–of–the–art approaches by investigating the reliability, the storage need and the robustness against hacker attacks. There might be further gains related to the coded distributed storage ideas if it comes to the delay in retrieving the data from distributed servers. As in BitTorrent, there might be a speed up in retrieving data from different sources. But even without the delay considerations, the other three parameters have shown the significant gain by using network coding and distributed storage.

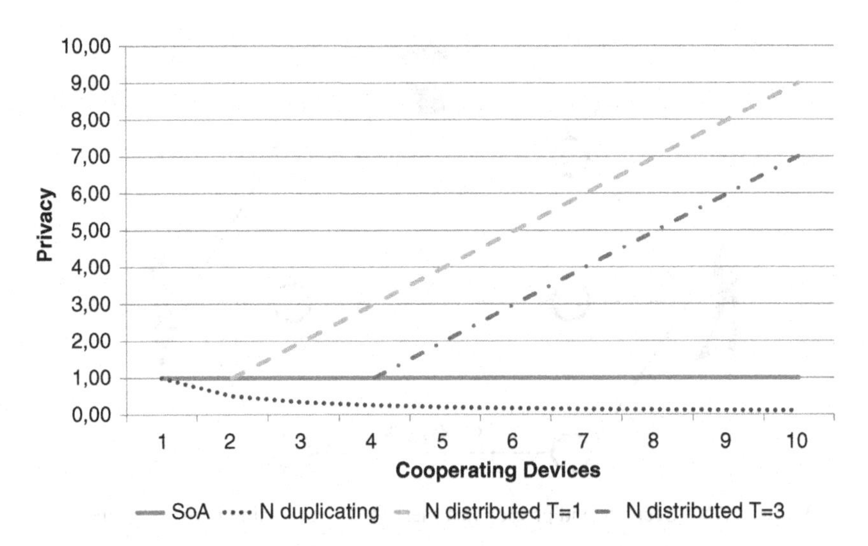

Figure 5.17 Robustness against hacker attacks for the three different approaches.

5.5.3 Security, Privacy and Data Integrity in Mobile Clouds

Security and privacy are always fundamental issues when it comes to mobile cloud services. As we cooperate with other users that we may know or may not known at all, the risk is that some users might monitor unintended data. This may not be a big issue for multicast services as this is intended to be received by multiple users but it may be for unicast services. As we have seen in the example of the distributed storage, network coding alone (maybe in addition with other security mechanisms) is sufficient to protect the data. For the distributed storage we claimed that the data should be secure as long as the data is not fully stored on one cooperating partner. The same applies for streaming services. As long as the incoming streams are relayed over a multiplicity of devices, each relay will only have partial information. If then the data is coded over all streams, a potential hacker would need the key and all relayed streams in order to see the full data. This makes potential attacks even harder than attacks on state–of–the–art communications systems that are mainly dominated by point–to–point communications or single–path communications.

Figure 5.18 shows a potential combination of multi–path communication and network coding. Here we have three routes between two users and on each of the routes a linear combination of packets is transmitted. The potential intruder in the example is just overhearing one stream and by no means would that give the attacker any information. If just diversity would have been applied the hacker could retrieve one third of the raw data.

Another form of attack is to manipulate the incoming streams in order to destroy an ongoing communication. Here the main intention is not to overhear ongoing communication but at least make sure to destroy it.

Figure 5.19 depicts the scenario for a potential attack. The problem of intentional insertion of malicious data can be tackled by homomorphic coding [30, 31] in order to maintain data

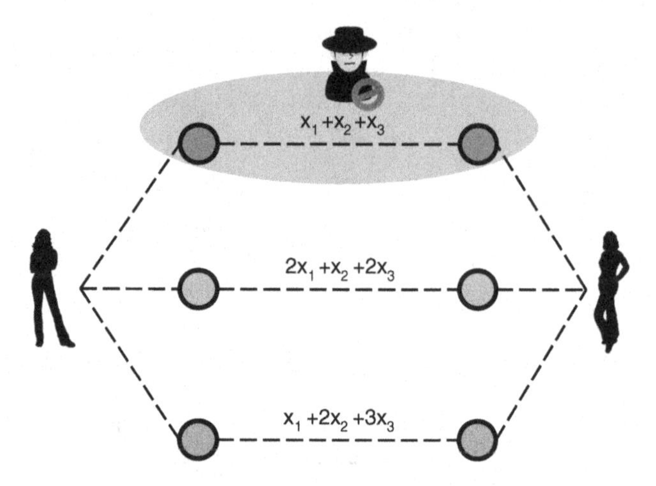

Figure 5.18 Security and privacy in a meshed network using network coding.

integrity. As a single coded packet has no context anymore and the bits are totally random between sender and receiver, an intruder could try to introduce random data just to destroy the performance of the network coding enabled communication. Packets that are coded from the original source have some kind of unique fingerprint that allows the receiver to identify whether they came from the original source or whether a potential intruder tried to send some data if network coding is used. Interested readers are referred to [30, 31].

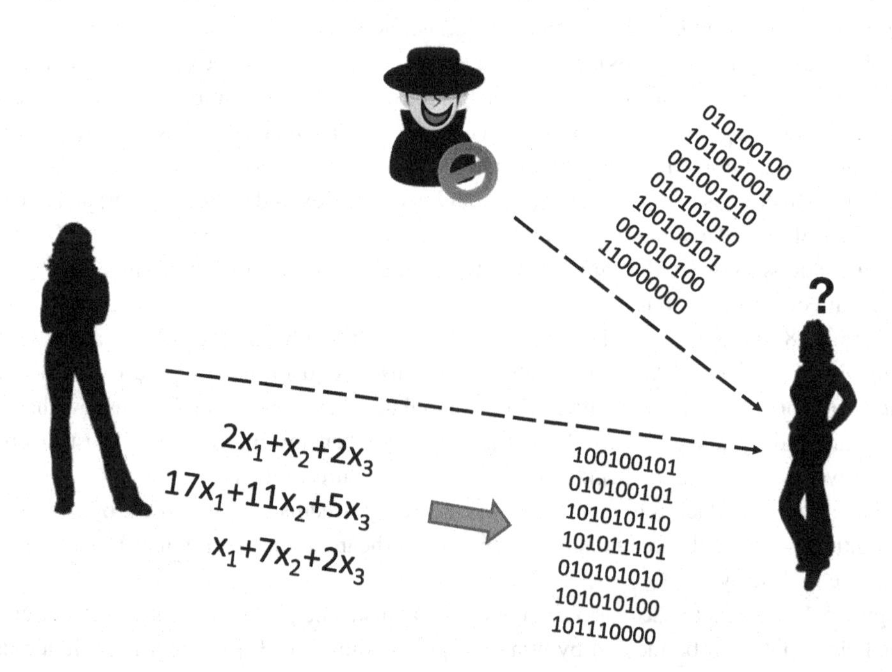

Figure 5.19 Data integrity in a network coding scenario.

5.6 Conclusion

Network coding is a key technology for mobile clouds as it increases the throughput of meshed networks, inherently provides security and introduces flexibility in data storage. The impact of network coding on energy savings is also very significant as it reduces the number of transmissions towards and within a mobile cloud. With respect to the throughput increase in meshed networks, intra–flow network coding has several advantages over inter–flow network coding as it requires less organization among the mobile nodes. In the future, the two coding schemes will be combined in order to jointly exploit their advantages, as proposed in [32]. There is higher complexity involved in intra–flow coding, but it has been shown that current implementations are running at sufficiently high coding speeds on any embedded system [22]. In general, network coding assures that the maximal capacity of a given network can be achieved. Furthermore, we have shown in this chapter that relay nodes have reduced load if network coding is applied. This has a positive impact of the overall network capacity, but also helps to convince third party nodes to join cooperation from a social point of view. In this chapter, we have also shown how to use network coding for distributed storage on mobile devices. Furthermore the advantage of using network coding over state-of-the-art approaches has been demonstrated. Network coding will play a major role in future communication systems as it will help to boost mobile cloud operation.

References

[1] R. Ahlswede, M. Cia, S.Y.R Li and R.W. Yeung. Network Information Flow. *IEEE Transactions on Information Theory*, 46(4):1204–1216, 2000.

[2] D.A. Patterson, G. Gibson and R.H. Katz. A Case for Redundant Arrays of Inexpensive Disks (raid). In ACM New York, editor, in *Proceedings of the 1988 ACM SIGMOD International Conference, on Management of Data*, pages 109–116, NY, USA, 1988.

[3] P. Elias, A. Feinstein and C.E. Shannon. Note on Maximum Flow through a Network. *IRE Transactions on Information Theory IT-2*, pages 117–119, 1956.

[4] S. Katti, H. Rahul, W. Hu, D. Katabi, M. Medard and J. Crowcroft. Xors in the Air: Practical Wireless Network Coding. In ACM Press, *In Proceedings of the 2006 Conference on Applications, technologies, architectures, and protocols for computer communications (SIGCOMM06)*, pages 243–254, 2006.

[5] M. Hundebøll, J. Leddet-Pedersen, J. Heide, M.V. Pedersen, S.A. Rein and F.H.P. Fitzek. Catwoman: Implementation and Performance Evaluation of IEEE 802.11 based Multi-Hop Networks using Network Coding. In *IEEE VTS Vehicular Technology Conference Proceedings*, IEEE, 2012.

[6] S. Gollakota and D. Katabi. Zigzag Decoding: Combating Hidden Terminals in Wireless Networks. In *ACM SIGCOMM 2008 Conference on Data communication (SIGCOMM '08)*, pages 159–170, New York, NY, USA, 2008.

[7] S. Katti, S. Gollakota and D. Katabi. Embracing Wireless Interference: Analog Network Coding. In *Applications, technologies, architectures and protocols for computer communications (SIGCOMM '07)*, pages 397–408, ACM, New York, NY, USA, 2007.

[8] S. Zhang, S.C. Liew and P.P. Lam. On the Synchronization of Physical-Layer Network Coding. In *IEEE Information Theory Workshop*, pages 404–408, 2006.

[9] F. Zhao and M. Medard. On Analyzing and Improving Cope Performance. In *Proc. of Info. Theory and App. Workshop (ITA)*, 2010.

[10] F. Zhao, M. Medard, M. Hundebøll, J. Ledet-Pedersen, S.A. Rein and F.H.P. Fitzek. Comparison of Analytical and Measured Performance Results on Network Coding in IEEE 802.11 Ad-Hoc Networks. In *The 2012 International Symposium on Network Coding*, June 2012.

[11] A. Paramanathan, U.W. Rasmussen, M. Hundebøll, S.A. Rein, F.H.P. Fitzek and G. Ertli. Energy Consumption Model and Measurement Results for Network Coding-Enabled IEEE 802.11 Meshed Wireless Networks. In

IEEE International Workshop on Computer-Aided Modeling Analysis and Design of Communication Links and Networks (CAMAD), Barcelona, Spain, 2012.

[12] K.F. Nielsen, T.K. Madsen and F.H.P. Fitzek. Network Coding Opportunities for Wireless Grids formed by Mobile Devices. In Springer, in the ICST Lecture Notes (LNICST) series, editor, *The Second International Conference on Networks for Grid Applications*. ICST, October 2008.

[13] K.F. Nielsen and F.H.P. Fitzek. Youtube video: Network coding n810. http://www.youtube.com/watch?v=VZYLSyZaEO8.

[14] B.A.T.M.A.N. – better approach to mobile ad–hoc networking. http://www.open-mesh.org/projects/batmand/wiki, 2007-2013.

[15] G. Ertli, A. Paramanathan, S. Rein, D. Lucani and F.H.P. Fitzek. Network Coding in the Bidirectional Cross: A Case Study for the System Throughput and Energy. In *IEEE VTC2013-Spring: Cooperative Communication, Distributed MIMO and Relaying*, Dresden, Germany, June 2013.

[16] M. Hundebøll, S.A. Rein and F.H.P. Fitzek. Impact of Network Coding on Delay and Throughput in Practical Wireless Chain Topologies. In *IEEE CCNC - Wireless Communication Track*, 2013.

[17] T. Ho, R. Koetter, M. Medard, D. Karger and M. Ros. The Benefits of Coding Over Routing in a Randomized Setting. In *Proceedings of the IEEE International Symposium on Information Theory, ISIT03*, 2003.

[18] F.H.P. Fitzek, J. Heide, M.V. Pedersen and M. Katz. Implementation of Network Coding for Social Mobile Clouds. *IEEE Signal Processing Magazine*, January 2013.

[19] J. Heide, M.V. Pedersen, F.H.P. Fitzek and T. Larsen. *Network Coding in the Real World*, chapter 4, pages 87–114. Academic Press, Oct 2011.

[20] M. Pedersen, J. Heide, F.H.P. Fitzek and T. Larsen. Network Coding for Mobile Devices - Systematic Binary Random Rateless Codes. In *Workshop on Cooperative Mobile Networks 2009 - ICC09*. IEEE, June 2009.

[21] M.V. Pedersen, J. Heide, F.H.P. Fitzek and T. Larsen. A Mobile Application Prototype using Network Coding. *European Transactions on Telecommunications (ETT)*, 21(8):738–749, December 2010.

[22] M.V. Pedersen, J. Heide, P. Vingelmann and F.H.P. Fitzek. Network coding over the $2^{32} - 5$ prime field. In *IEEE International Conference on Communications (ICC) Symposium*, Budapest, Hungary, June 2013.

[23] Q. Zhang, J. Heide, M.V. Pedersen, F.H.P. Fitzek, J. Lilleberg and K. Rikkinen. *Network Coding and User Cooperation for Streaming and Download Services in LTE Networks*, chapter 5, pages 115–140. Academic Press, Oct 2011.

[24] L. Militano, F.H.P. Fitzek, A. Iera and A. Molinaro. Evolutionary Theory for Cluster Head Election in Cooperative Clusters Implementing Network Coding. In *European Wireless 2009*, Aalborg, Denmark, May 2009.

[25] L. Militano, F.H.P. Fitzek, A. Iera and A. Molinaro. A Genetic Algorithm for Source Election in Cooperative Clusters Implementing Network Coding. In *IEEE International Conference on Communications (ICC 2010) - CoCoNet Workshop*, May 2010.

[26] L. Militano, F.H.P. Fitzek, A. Iera and A. Molinaro. Network Coding and Evolutionary Theory for Performance Enhancement in Wireless Cooperative Clusters. *European Transactions on Telecommunications (ETT)*, 21(8):725–737, December 2010.

[27] L. Militano, F.H.P. Fitzek, A. Iera and A. Molinaro. Data Seeding in Nomadic Cooperative Groups. In *Sixth Workshop on multiMedia Applications over Wireless Networks (MediaWiN) in association with the Sixteenth IEEE Symposium on Computers and Communications (ISCC 2011), Kerkyra, Greece*, 2011.

[28] L. Militano, F.H.P. Fitzek, A. Iera and A. Molinaro. Group Interactions in Wireless Cooperative Networks. In *Wireless Access, IEEE Vehicular Technology Conference (VTC) - Spring 2011*, Budapest, Hungary, 15-18 May 2011. IEEE.

[29] P. Vingelmann, M.V. Pedersen, F.H.P. Fitzek and J. Heide. Data Dissemination in the Wild: A Testbed for High-Mobility MANETS. In *IEEE ICC 2012 - Ad-hoc and Sensor Networking Symposium*, June 2012.

[30] C. Gkantsidis and P. Rodriguez. Cooperative Security for Network Coding File Distribution. In *IEEE Infocom*, 2006.

[31] Maxwell and N. Krohn. On–the–fly Verification of Rateless Erasure Codes for Efficient Content Distribution. In *Proceedings of the IEEE Symposium on Security and Privacy*, pages 226–240, 2004.

[32] J. Krigslund, J. Hansen, M. Hundebøll, D. Lucani and F.H.P. Fitzek. Core: Cope with More in Wireless Meshed Networks. In *IEEE VTC2013-Spring: Cooperative Communication, Distributed MIMO and Relaying*, Dresden, Germany, June 2013.

6

Mobile Cloud Formation and Maintenance

Nature is a mutable cloud which is always and never the same.

Ralph Waldo Emerson

This chapter discusses the dynamics of mobile clouds, particularly the situations where clouds experience changes affecting their operation. The following three mobile cloud operation stages can be identified: cloud formation, cloud operation and cloud maintenance. The first and last stages are the most relevant concerning dynamic changes, while the second stage assumes that the cloud remains unchanged during that period. In addition to studying cloud formation and maintenance stages, this chapter also discusses possible methods that can be used to manage the changes in the cloud. Service discovery, deeply related with cloud formation and maintenance, is also considered in this chapter.

6.1 Introduction

Mobile devices, the nodes of a mobile cloud, freely move and roam, and hence a cloud is in principle subject to dynamic changes during its operation. Changes include mobile devices joining or leaving the cloud, as well as devices moving within it, altering the connectivity between nodes and therefore changing the cloud topology. Changes also take place due to fluctuations in the availability and instantaneous conditions of resources residing in the nodes. A key issue that needs to be understood is how to manage these changes in an efficient way. This chapter discusses this issue, taking into consideration the particular characteristics of mobile clouds. Even though there is a considerable body of research on the subject for mobile ad hoc networks (MANET), the presence of centralized entities, e.g.,

Mobile Clouds: Exploiting Distributed Resources in Wireless, Mobile and Social Networks, First Edition.
Frank H.P. Fitzek and Marcos D. Katz.

base stations and access points brings a new perspective into managing the dynamics of a mobile cloud.

6.2 Mobile Cloud Stages

A mobile cloud is, by definition, a dynamic system where nodes are characterized by their mobility. It follows that interactions among nodes are in principle opportunistic. There are many aspects related to how a mobile cloud is dynamic. First, the formation of the mobile cloud itself. Then, once the mobile cloud has reached a given generic size N, one can expect that there will be subsequent changes in the cloud, due to mobility of the nodes, fluctuations in the state of nodes resources and decisions made by node users, for instance. In general we can define three basic stages in the operation of a mobile cloud, namely:

Mobile Cloud Formation: this is the initial process from which a mobile cloud progressively grows from a single–node (no cloud) to a cloud of size N. This process is illustrated on Figure 6.1a. One may wonder, what makes nodes to join a mobile cloud, how this operation is managed in practice, and many other related questions. At this point we just refer to a service discovery procedure, which broadcasts locally, that is in the operating area of the nodes, information about services being provided in that area, and other possible service information. Service discovery will be described in more detail later in this chapter. A new node may decide to join a particular service being currently used by other nodes, already members of a mobile cloud. In the most general case, nodes, or more precisely, users controlling them, need a clear encouraging incentive to join a mobile cloud, as typically people would expect to obtain certain benefits out of their cooperative action. Incentives can be purely technical, such as enhanced data support and QoS, more efficient use of the battery, and others. Certainly, incentives can also take place or be offered in other domains, such a financial and social.

Mobile Cloud Operation: in this stage the mobile cloud is assumed to be in a stable condition, in terms of number of participating nodes, node relationships (e.g., distance) and state of the node resources, see Figure 6.1b. This stage represents the main assumption made when devising cooperative strategies for, and analyzing the operation of, mobile clouds. In real life such a stable condition exists for a given period of time, depending on dynamic nature of the nodes and operating scenarios. Since this chapter addresses mostly the dynamic aspects of mobile clouds this particular stage will not be discussed in further detail. Note that other chapters dealing with cooperation between nodes of the cloud and between nodes and overlay network assume operation in this stage.

Mobile Cloud Maintenance: in practice, and concerning mobility, the nodes of a mobile cloud can be a mixed combination of mobile, movable and fixed devices. New nodes could join an operating cloud, and already engaged nodes could leave from it, at any time. In addition, nodes can move within the cloud,

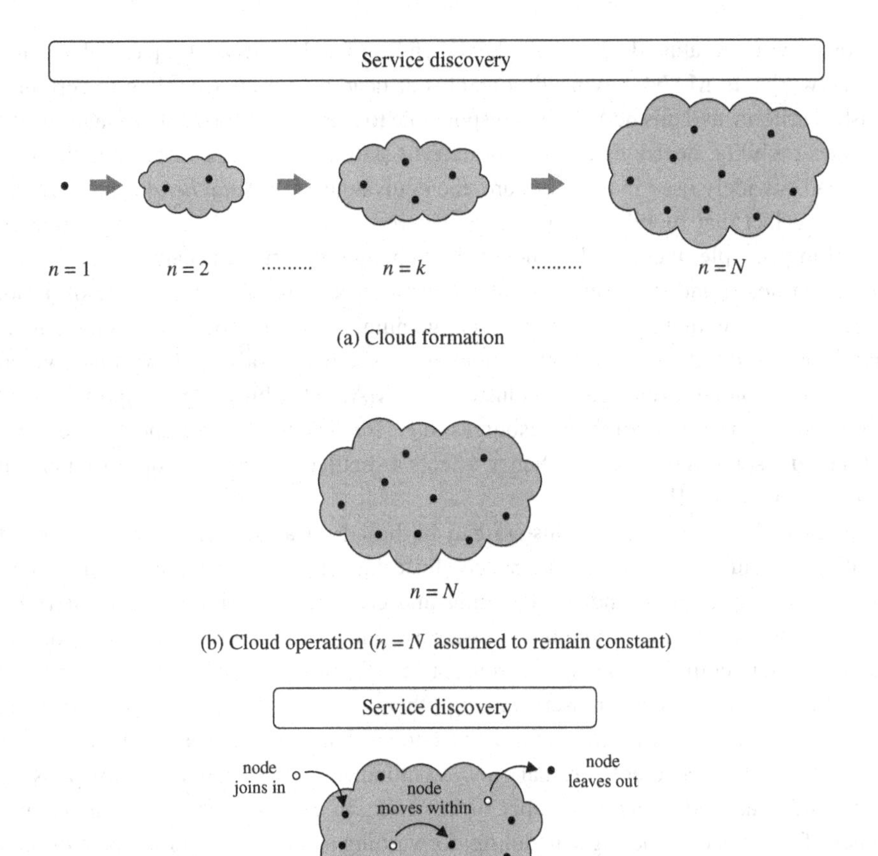

Figure 6.1 Stages of mobile clouds: a) cloud forming, b) cloud operation, and c) cloud maintenance.

possibly changing the connectivity conditions between them. Figure 6.1c depicts these typical cloud maintenance situations. There are other changing conditions in the cloud that need to be managed, including a) decisions made by users on the use of their mobile devices and services, b) changes in the availability and conditions of both, individual (e.g., on the device) and shared (e.g., common) resources, c) selfish user behavior and d) security threats.

At this point one may wonder how cloud formation and maintenance is managed, and which are the entities involved in these procedures? In general we refer to mobile cloud management to the procedures involving both formation and maintenance of the cloud. Let us first approach cloud management from the component networks of a mobile cloud. On one hand the mobile cloud can be seen as an ad hoc network of interacting mobile nodes. From this perspective there exists a vast literature dealing with the dynamics of ad hoc networks, particularly mobile ad hoc networks. In MANET typically one cluster head is chosen according to certain pre–established criteria, and this node is then responsible to manage any possible changes resulting from nodes mobility, decisions of users or current state of resources. Even though the term cluster head is widely used in the literature, the equivalent term *cloud head* is preferred here, to reflect the fact that mobile clouds are considered. There could be more than one cloud head, and in principle, these leading nodes may change with time, because of the changing conditions in nodes and the cloud, and also due to fairness reasons. If the size of a mobile cloud becomes large, managing efficiently a large number of nodes becomes a challenge, and it makes sense to divide the cloud into a number of smaller clouds, each with its own cloud head. This partitioning is referred to as clustering in MANET. Clustering is important because a) it facilitates the reuse of common resources (e.g., frequency), b) it enhances the utilization of distributed resources (e.g., every change affects a smaller number of nodes), c) it enhances the routing process [1, 2].

Grouping nodes into different clusters can be based on a number of criteria, including dominating–set clustering, low–maintenance clustering, mobility–aware clustering, energy–efficient clustering, load–balancing clustering and combined metrics–based–clustering [2]. From the standpoint of mobile clouds, clustering can also be carried out based on users' interests to certain common contents, affinity or complementarity of resources on the nodes, and social relationships between users, among others. In general three types of nodes can be defined in a cluster, namely normal (or member) nodes, cluster/cloud head nodes and gateway nodes [3]. Normal nodes refer to ordinary nodes without other particular cluster responsibility. Cluster/cloud head nodes have a number of key tasks to carry out, including managing the dynamics of the cluster, handling scheduling, forwarding and routing operations in the cluster and performing power control, acting just as a temporary access point within the cluster. Gateway nodes serve as connectivity links between clusters. Clustering as such is beyond the scope of this chapter, but interested readers are referred to [1–6] for a comprehensive discussion on this subject.

The dynamics of a mobile cloud could also be managed by the other component network, that is the overlay cellular network. In this approach, any possible change in a node needs to be informed to the controlling base station. We assumed that there is an entity on the network side that is able to receive, update and manage information of the nodes. A dedicated Cooperative Control Server (CCS) has been proposed in the literature for that purpose [7, 8]. The role of the CCS is in principle similar to that of a cluster or cloud head, but of course that entity is now located far away from the cloud, on the network side. The key advantage of the CCS is its central access to each of the mobile nodes of the cloud. We can say that CCS oversees the whole cloud from above, in a direct and simple manner. Even though a cloud head will do the same task, the head itself is much more prone to the effects of the dynamic and opportunistic characteristics of the cloud. For instance, if the cloud head leaves the cloud or is switched off, it needs to previously transfer all its cloud–related information to a newly selected cloud. This could be challenging as the change should happen in a transparent manner to all nodes of the cloud. As compared with the local managements of the cloud, accessing the overlay network is slow as well as expensive in terms of energy and spectrum usage. Indeed, every node connected to a base station will exchange information with the CCS on a regular or event–driven basis, consuming for that purpose licensed cellular spectrum, and a relatively high amount of energy. Any important change on the cloud will always be registered on the CCS, which, since it resides outside the cloud, is unaffected by topological or any other changes in the cloud.

One of the most important issues when considering the management of a mobile cloud is the performance and cost of the solution. Performance refers mostly to the adaptation speed as well as to the impact on the quality–of–service perceived by the users. Cost refers to several issues, like the required amount of common and node resources consumed to form and maintain a mobile cloud (e.g., spectral and energy expenditures, respectively), complexity (e.g., overhead and implementation requirements). Mobile clouds combine a centralized and distributed architecture, and it is precisely the hybrid topology that allows managing of the cloud efficiently, aiming at exploiting the best of both worlds. The concept of cloud leader and CCS can be naturally merged in a mobile cloud, resulting in a system with both a local and a remote managing approach. High speed of reaction, flexibility and efficient use of energy and spectral resources result from the local management, while the CCS provides a simple centralized approach that is robust to dynamic changes in the cloud. Figure 6.2 illustrates the three described architectural approaches to manage a mobile cloud, namely, a) locally (pure ad hoc), b) remotely (pure cellular), and c) in a distributed or hybrid (cellular and ad hoc) fashion.

Figure 6.2c shows one possible way to implement a distributed management, establishing a direct link between the cloud head and the CCS. This simple approach, however, does not exploit the inherent diversity available in mobile clouds. For instance the multiplicity of nodes readily brings redundancy into the system, and multiple parallel uplink or downlink connections can be established between the nodes and the base station or access point to enhance the performance and quality of the wireless connections. Of course, this is made at the cost of increasing signaling and energy consumption.

In order to get a better insight on the performance of different architectural solutions for mobile cloud management, we compare four different approaches, in terms of energy required to form a cloud of size N as well as the latency involved in that process. It is assumed that the cloud is formed through a simple service discovery procedure such that each mobile device broadcasts a *Hello*–message containing its identification and information on services it is interested in. After that the mobile goes into a fixed–time listen–state, receiving similar information from neighbor peers. This procedure is repeated regularly to take into account dynamic changes in the cloud. We consider only devices joining the cloud as we are here interested in studying the performance of cloud formation. Only one–hop connections between neighbor mobile devices are assumed. We consider here five architectures for managing the cloud, namely a) local (ad hoc), Figure 6.2a, b) remote (cellular), Figure 6.2b, c) distributed Hybrid Uplink and Downlink, (HUD), Figure 6.2c, d) distributed Cellular Uplink Hybrid Downlink (CUHD), and e) distributed Hybrid Uplink Cellular Downlink (HUCD).

CUHD corresponds to a hybrid case where uplink management signaling is carried out by cellular links (cellular to CCS) and downlink signaling is carried out using a direct link between CCS and the mobile head. In the same way, HUCD represents the case where uplink signals are transmitted from the cloud head to the CCS, while downlink signals are individually

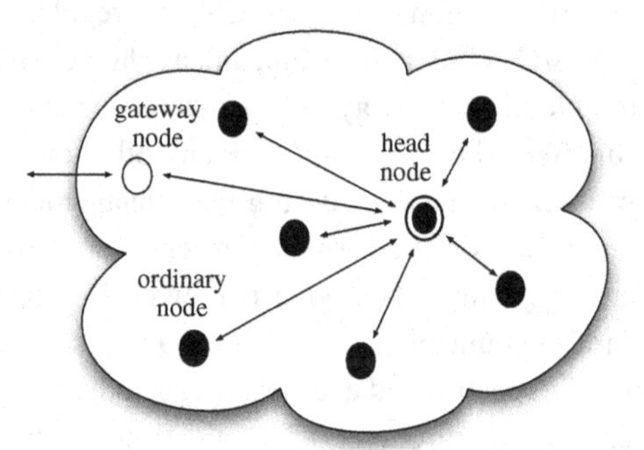

(a) Local management of a mobile cloud

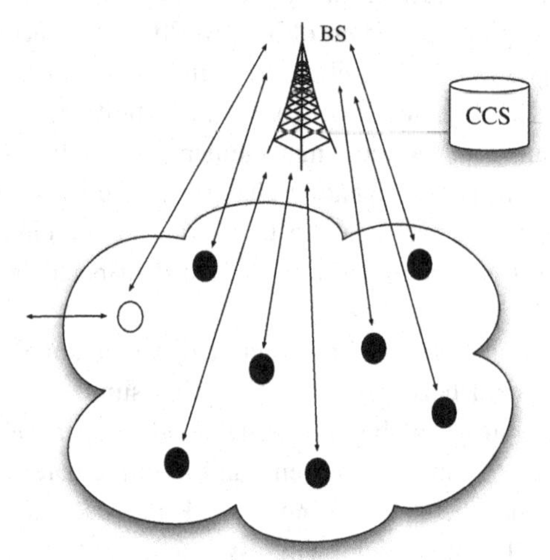

(b) Remote management of a mobile cloud

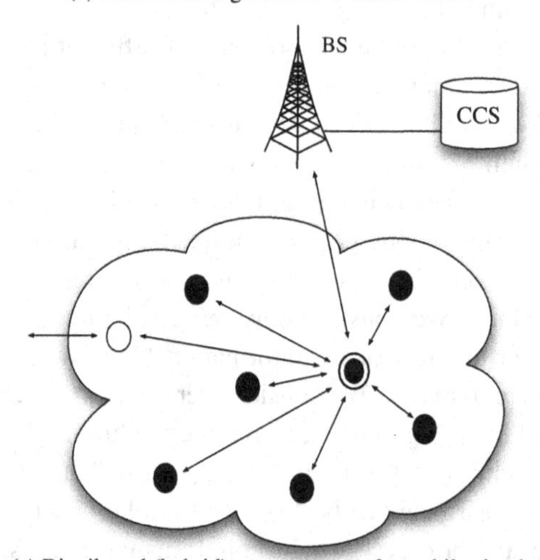

(c) Distributed (hybrid) management of a mobile cloud

Figure 6.2 Architecture approaches for managing a mobile cloud: a) local (ad hoc), b) remote (cellular) and c) hybrid.

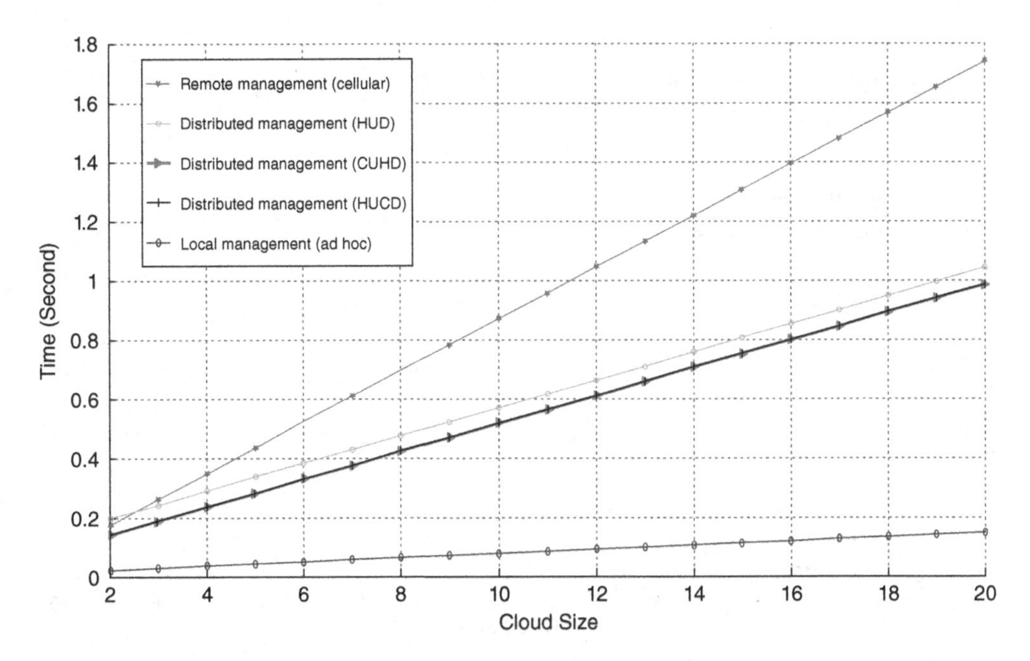

Figure 6.3 Time required to form a mobile cloud of size N for different management architectures.

transmitted using N cellular links. Figure 6.3 shows the time required to form a cloud of size N, $N = 2$–20, assuming that the N devices are joining one after the other, without additional time between them other than the time required by the service discovery protocol. Additional details of the models and assumptions can be found in [7]. As expected, the fastest response results from the case where the cloud is locally administered (ad hoc), while the slowest case corresponds to the remote management approach. Note that local interactions result in shorter processing delays due to physical closeness and because links between devices in the cloud support higher throughput than do cellular links. The delay performance of the distributed approach falls between the two extreme cases.

For small clouds, say $N < 20$, the overall processing delay in not highly significant. Figure 6.4 illustrates the energy expenditure needed to form a cloud of size N for the five schemes under consideration. We can see again a somewhat similar behavior, whereby local management is the most energy efficient, as expected, whereas remote management consumes the most energy. However, the performance of the distributed approach connecting directly the cloud leader with the CCS through a single link (HUD) is close to that of the local approach. With this example we illustrate the impact of the cloud management architecture on the performance. However, we stress the fact that there are other issues that need to be taken into account with assessing the overall performance. The impact on performance of the dynamics of the nodes is important. Even though the purely ad hoc approach appears to be the most attractive scheme (based on Figures 6.3 and 6.4), a distributed cloud managing approach can provide a more reliable architecture, exploiting the fact that the overlay network is stable and continuously operating, in comparison with the dynamic nodes.

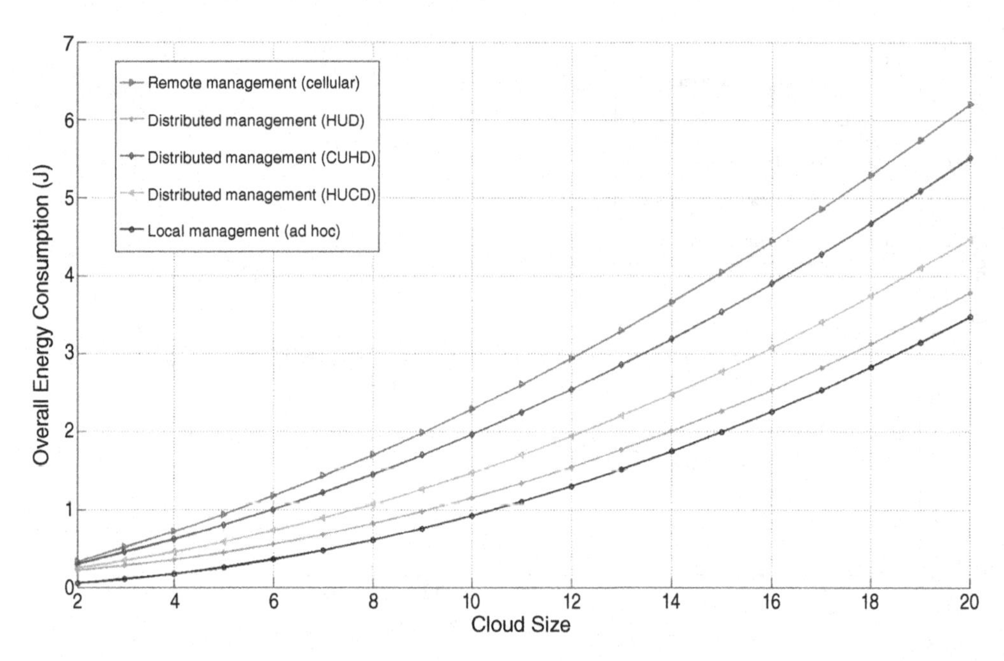

Figure 6.4 Overall energy expenditure to form a mobile cloud of size N for different management architectures.

6.3 Service Discovery for Mobile Clouds

This section discusses service discovery for mobile clouds in a very introductory manner. Service discovery is in general defined as the process supporting automatic detection of devices and services. This section focuses on service discovery particularly designed for mobile clouds. Current solution such as Zeroconf [9] and Bonjour (Apple's implementation of Zeroconf) are not described here but are explained later. The activities of Alljoyn are discussed in Chapter 10.

The very first question to be answered is: how can independent mobile devices be aware of certain cooperative services that they could be jointly using? The question can be answered in many ways, as the service can be originated at different locations, within the mobile cloud, or outside it. Moreover, the answer also depends on the entity or entities providing the service and the type of service being provided. Given the inherent flexibility and versatility of mobile clouds, and considering their capabilities as a platform for social interaction, the term service provision used in the context of mobile clouds is broader than in the conventional case, where typically one or more direct service requesters (e.g., clients) and a single service provider are considered.

The concept of service has evolved in the last decades and new service paradigms have emerged, mostly driven by the developments in networking and device technology. Mobile services conventionally have been provided by network operators and service providers, typically using a centralized client–provider model. Certainly service in distributed ad hoc networks has been also a very important area of development, where one or mode nodes can host a certain service. The fact that mobile devices are equipped with increasingly powerful communications, processing and sensorial systems has made it possible that a user can easily become a content provider and ultimately a service provider. Large amounts of information stored in and produced by a mobile device (e.g., live video) can be readily traded to other users, mobile or not, based on different service principles. This can be done following different models, from merely sharing the information among friends to delivering a requested service based on a financial reward. These ideas can be extended to the case of a mobile cloud, where users on a cloud agree to cooperate to produce a particular service. Such a mobile cloud service could be based on the contributions of some or all devices forming the cloud. By this, the devices put some designated resources into a resource pool, from which the service is created. Such a service can be offered to members of the mobile cloud, or eventually to anyone interested, outside the cloud. For instance mobile cloud users may combine microphones and image sensors to improve the quality of the captured sound and video. This action can be designed not only to be enjoyed by the cloud members themselves, but also a profit–making goal could be the target in mind.

Figure 6.5 illustrates the wide array of possible service providers as well as clients in the field of mobile clouds. One, few or all nodes of the mobile cloud can provide a given service (in–the–cloud service provision) on one hand, or the service could be given by a network

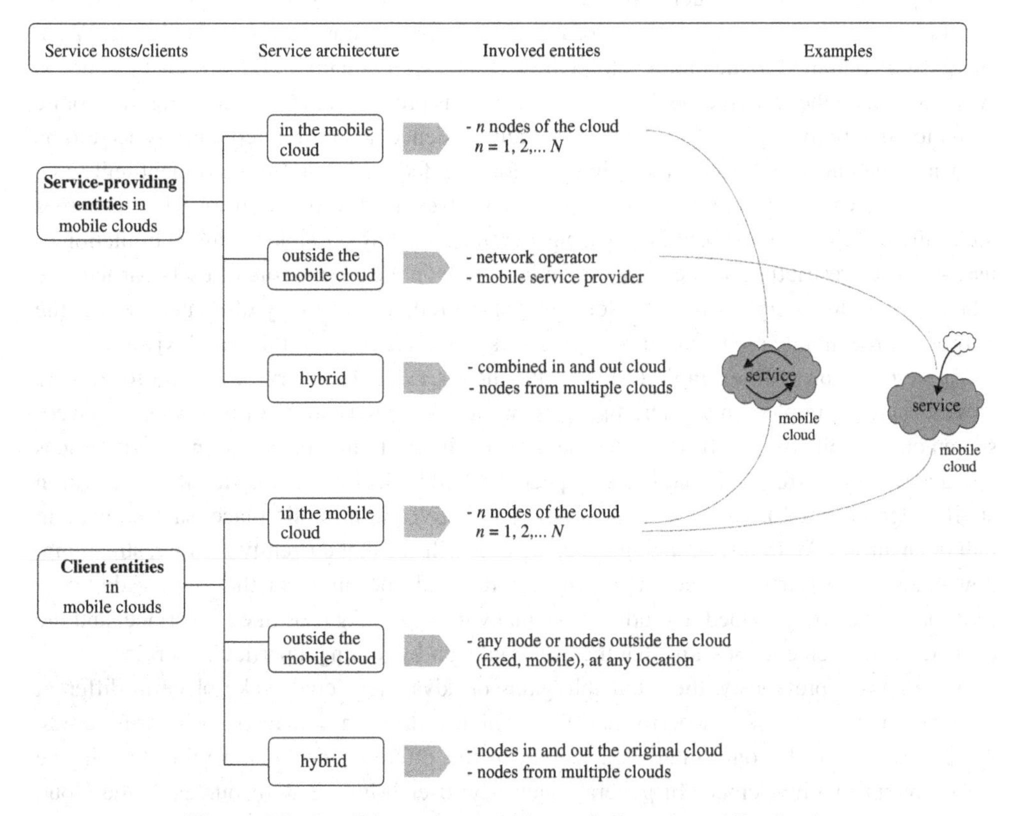

Figure 6.5 Possible service providers and clients in mobile clouds.

operator or mobile service provider (out–of–cloud service provision), on the other hand. In the former case nodes cooperatively share resources in order create a service. Hybrid cases can also be considered, where both cloud nodes and external entities jointly create a service. In addition to the service hosts, it is important to identify the service targets, or clients. These can also be nodes members or the mobile cloud, out–of–cloud nodes as well as both targets simultaneously, as shown in Figure 6.5. Note that the cases depicted in Figure 6.5 only identify service-providing entities and their counterpart service recipient entities, while service discovery architecture and management are not included on the figure.

Two examples of possible service provision and clients for mobile clouds are shown in Figure 6.5. In one case, far right on Figure 6.5, the service providers reside outside the mobile cloud while the clients are in the cloud. This could be in practice a case of cooperative content delivery, like those discussed in Chapter 9, where users in the cloud are interested in the same service, e.g., receiving a real–time streaming video from a live event. Certainly the service provider is outside the mobile cloud and clients are the nodes of the cloud. The other example in Figure 6.5 considers the case where service providers and clients all belong to the mobile cloud. Let us consider here the example where users share their loudspeakers to create a three-dimensional sound-processing platform. Obviously, the users themselves are both service providers and clients of the service.

As previously discussed, each mobile device has a number of resources onboard, which can be real physical resources, such a processing power, sensors, mass memory and multiple air interfaces, or intangible ones, such as information or applications stored on the device. When shared and combined in the mobile cloud these can be seen as a pool of distributed resources. A discussion on the resources and their possible uses is presented in Chapter 3. For the mobile cloud to potentially exploit these resources, their existence and current conditions need to be known to the entity or entities managing the cloud. In fact, the cloud-management unit (e.g., cloud leader, CCS) and ultimately the users themselves need to be aware of what resources are available, how they are being used at a given time, and any possible technical limitation or user–imposed restriction on their usage. We lay stress on the crucial role of users particularly when service discovery for mobile clouds is considered, as users may ultimately decide the purpose of the mobile cloud based, among others, on resource availability and type.

Figure 6.6 depicts an example of how a practical service discovery procedure for mobile clouds could be implemented. The figure shows a conceptual model of a service discovery screen on a mobile device, from where the user obtains real–time information of the services being currently offered and used by other peers located in the immediate vicinity (e.g., within short–range coverage). Let us assume a new user arrives in a public place, say the waiting hall of an airport. With just a quick glance, the user will see that currently gaming, streaming, download, and storage services are being offered and consumed in that area. Additional information can be provided, including how many users are engaged in each service, and the expected enhancements or gains that the user could get by joining a particular service.

As discussed previously, the attainable gains or advantages could take place in different domains, more precisely, in performance, resource utilization and resource–sharing areas. A new user could become interested in one of the offered services, deciding to join the corresponding mobile cloud. In general, each new user brings new resources to the cloud

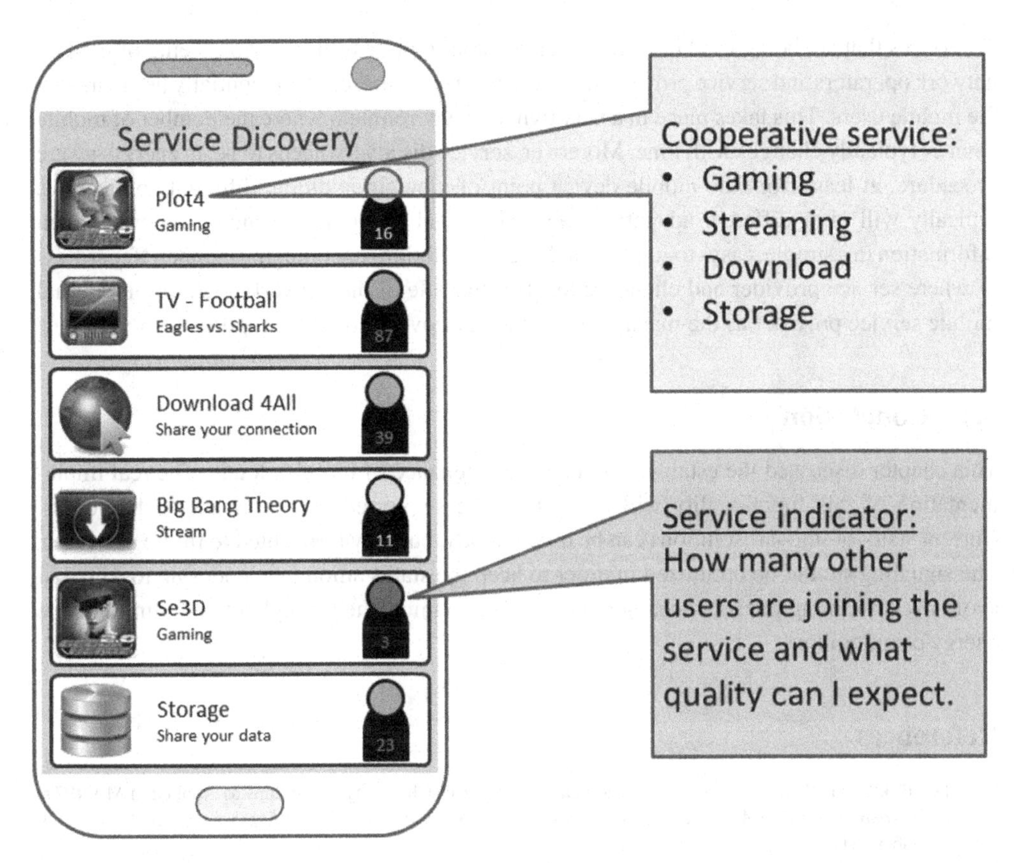

Figure 6.6 Example of a possible service discovery for mobile clouds showing the available services.

and this would, in principle, benefit all members of the cloud. Several examples are shown in [7, 8], where different strategies for content delivery in mobile clouds are studied. It is shown that the more users in the cloud, the less energy is required to carry out the task by each mobile device of the cloud. A new user may find no particular interest in joining any of the available services published by the service discovery procedure, preferring instead to join another service, not yet offered within that area. By doing this, the user not only brings a new service to the area but also creates the basic infrastructure of a possible future mobile cloud. Indeed, the mobile device of the new user is the initial node ($N=1$) of a mobile cloud, and eventually other devices will join in. Once this user initiates a new service, the service discovery procedure will broadcast information on it.

Service discovery is essential for mobile clouds, as it is through this procedure that mobile clouds come into existence. Users will cluster and form a mobile cloud based on the service advertisements and further guidance of an ongoing service discovery procedure. An eclectic array of possible applications can be created around the concept of mobile cloud, and in principle the service discovery procedure needs to be take into account many of these possible applications. This is one of the key challenges for service discovery. Users need to be informed of services that are being used in a given area by other mobile users, services being offered by network operators and service providers, and other services that can be potentially be created by the mobile users. This takes place in a very dynamic environment where the number of mobile devices typically changes with time. Moreover, service discovery needs to be an energy–aware procedure, at least from the mobile device point of view. In addition, different applications typically will offer different advantages and gains, and it could be difficult to convey such information in a simple, easy–to–understand and cooperation–encouraging manner. Regardless of where service provider and client are located, the role of the network operator or external mobile service provider as the manager of service discovery could be important.

6.4 Conclusion

This chapter discussed the establishment and maintenance of mobile clouds. The real implementation of this functionality will depend on the supported use cases and technologies. Current state–of–the–art solutions can be used already, but may be adapted to future use cases. The signaling should be optimized in order to keep the status among the nodes up to date, but avoid overwhelming packet exchange in order to maximize the throughout and minimize the energy consumption.

References

[1] M.U. Bokhari, H.S.A. Hamatta and S.T. Siddigui. A Review of Clustering Algorithms as Applied in MANETs. *International Journal of Advanced Research in Computer Science and Software Engineering*, 2(11):364–269, November 2012.

[2] J.Y. Yu and P.H.J. Chong. A Survey Of Clustering Schemes For Mobile Ad Hoc Networks. *IEEE Communications Surveys & Tutorials*, 7(1):32–48, 2005.

[3] B.A. Correa, L. Ospina and R.C. Hincapi. Survey of Clustering Techniques for Mobile ad hoc Networks. *Revista Facultad de Ingeniera Universidad de Antioquia*, pages 145–161, September 2007.

[4] P. Ghosh, S. Nandy, N. Pandey and M.K. Naskar. Energy aware Algorithm for Clustering in Wireless Network. *International Journal of Computer Applications*, 53(10):1–10, September 2012. Published by Foundation of Computer Science, New York, USA.

[5] I.G. Shayeb, A.H. Hussein and A.B. Nasoura. Survey of Clustering Schemes for Mobile Ad-Hoc Network (MANET). *American Journal of Scientific Research*, 1(20):135–151, 2011.

[6] Dali Wei and H.A. Chan. Clustering Ad Hoc Networks: Schemes and Classifications. In *3rd Annual IEEE Communications Society on Sensor and Ad Hoc Communications and Networks*, volume 3, pages 920–926, 2006.

[7] H. Bagheri, P. Karunakaran, K. Ghaboosi, T. Braysy and M. Katz. Mobile Clouds: Comparative Study of Architectures and Formation Mechanisms. In *IEEE 8th International Conference on Wireless and Mobile Computing, Networking and Communications (WiMob)*, pages 792–798, 2012.

[8] T. Chen and M. Katz. Cooperative Architecture for Cellular-short-range Combined Mesh Networks. In *Proceedings of the 5th International ICST Mobile Multimedia Communications Conference*, London, UK, 2009.

[9] D.H. Steinberg and S. Cheshire. *Zero Configuration Networking: The Definitive Guide*. O'Reilly, December 2005.

Part Three

Social Aspects of Mobile Clouds

7

Cooperative Principles by Nature

No instinct has been produced for the exclusive good of other animals, but each animal takes advantage of the instincts of others!

Charles Darwin (1859)

Cooperation is the process of working and interacting together aiming to achieve individual or common goals. Cooperation is present in our daily life, and ultimately, it is the basis of our society. Cooperation arises not only among humans but it is a widespread phenomenon in nature. Human cooperation is governed by intricate rules and patterns of behavior, much more complex that the cooperation arising in other species. Cooperation in the animal kingdom, being somewhat simpler though highly efficient, has often served as an inspiration source for many developments in engineering. Many cooperative strategies in the animal world can be particularly useful when applied to a group of user–controlled nodes (i.e., mobile devices) such as those in a mobile cloud. This chapter considers, through several examples, cases of cooperation in Nature, and how these can be used to derive basic rules of cooperation. The goal of this chapter is to motivate the reader to identify and exploit cooperative strategies for mobile clouds by finding some connections between cooperation found in Nature and cooperation arising among a group of cooperative mobile users. This chapter also discusses game theoretical approaches particularly motivated by the influential work of Robert Axelrod.

7.1 Introduction

In this chapter we present and discuss the principal concepts of cooperation based on examples found in Nature. Since the 1960's a great deal of research has been carried out in this field, but here we will try to understand key essential principles of cooperation. These principles,

Mobile Clouds: Exploiting Distributed Resources in Wireless, Mobile and Social Networks, First Edition.
Frank H.P. Fitzek and Marcos D. Katz.

inferred from cooperative behavior found in Nature, ultimately serve as inspiration sources while developing cooperative rules for mobile clouds. Understanding the cooperation rules of vampire bats, monkeys, cheetahs, and hyenas the most important rules can be derived. There are many more examples from Nature such the symbiosis of the lycaenid butterfly larvae with ants [1], and others. The rules are often very simple and can be directly implemented on mobile devices, leading to cooperative gains without any complexity. Human cooperation is excluded here, as cooperative behavior or the lack of it is governed by complex individual and social rules, often acting in many realms. For instance, people do not always optimize their cooperation behavior in order to get the highest possible gain (profit maximization), but they also consider other issues, such as ethical reasons.

For the impatient reader we list here the most important basic rules of cooperation with associated examples of cooperative behavior in the animal kingdom:

1. Reciprocity (vampire bat)
2. Detection and punishment of cheaters (vampire bat)
3. Invest into future cooperation (monkey)
4. Tolerance in pay–off (vampire bat and monkey)
5. Cognition of cooperation partners (vampire bat and monkey)

In order to understand the roots of these cooperation rules, the animals and their cooperative behavior are briefly described in the following sections.

7.2 Cheetahs and Hyenas

To underline the importance of cooperation we compare two animals with very different approaches towards cooperation. On one hand we consider the cheetah. The cheetah (*Acinonyx jubatus*) is the fastest animal on land, reaching velocities above 100km/h. The cheetah hunts mammals such as gazelles, impalas, and springboks with a success rate of approximately 50% [2]. In case the hunting was successful the cheetah needs to rest for some time to recover from the strain. In this time the cheetah risks that other animals like lions but most likely hyenas come to steal the prey animal. Even though a single hyena is body–wise smaller than a cheetah, the cheetah is simply outnumbered by several cooperating hyenas. The cheetah evolved over time to become slim and fast, enabling this species to hunt successfully more mammals. But considering situations with the hyenas, a slim body becomes a problem as it cannot defend the food, leading to the *evolutionary trap*. The hyenas also confront and attack lions [3] but for this the hyenas need a larger group and inexperienced lions. Recent research by Duke University shows that hyenas outperform even primates in cooperative problem–solving tests [4].

As already stated by Darwin, cooperation is a key feature for evolution [5]. Even though this example does not result directly into a cooperation rule, the example of the hyenas and cheetah underlines the benefit of cooperation (hyenas) over monolithic approaches (cheetah).

7.3 Orca – Killer Whales

Several animals are known for cooperative hunting strategies. Lions and hyenas cooperate just to outnumber the target, a simple yet effective strategy. A special strategy for cooperative hunting is used by orcas (*Orcinus orca*), commonly known as killer whales, hunting for weddell seals resting on ice floes. As a single orca is not able to reach the seal on the ice floe alone by either climbing on the ice floe or by pushing the floe over, the orca hunting strategy is based on cooperation among multiple individuals.

In [6] the highly sophisticated hunting strategy was reported for the first time with reliable research data. Even though it was spotted already in the 1970's there was no reliable evidence for this behavior of the orcas. The hunting strategy reported in [6] was based on several steps. First orcas would push the ice floe out into the open sea to isolate the ice floe from other surrounding ice floes. Then one orca referred here as the master orca would stay close to the ice floe and orchestrate the other orcas to execute with perfect synchronism a large wave to wash the seal from the ice floe. In order to create such a wave the orcas need to swim very fast underwater towards the ice floe. Just before they reach the ice floe the master orca emits tics and whistles to inform the cooperation group when to break the water (reaching the surface of the water and beating with the tail of their body on the water). It is clear that in order to create a wave with maximal impact the orcas need to isolate the target ice flow and bring it to the open sea as other floes would just disturb the perfect wave. Furthermore it is also beneficial for the orcas as the detached seal has no chance to climb other floes nearby and escape. With this sophisticated strategy orcas have a high success rate in hunting down the isolated seal.

Other techniques for cooperative feeding has also been reported for humpback whales and orcas. A group of whales cooperatively herded herring enable other whales to easily catch the herring. This is known as bubble–net feeding [7] or the carousel method [8]. As cooperative hunting is common among several species, the notable aspect of this hunting strategy is the planning and the controlling of the cooperation group by a single orca or group of whales. Also notable is the pay–off tolerance in the feeding gain as not all participating whales gain at the same time. In the following chapters we will see the same technique often referred to *cluster head*.

7.4 Vampire Bats

The vampire bat (*Desmondus rotundus*) provides a very illustrative example of cooperation. The reason is that the set of rules established by vampire bats is limited and clearly defined. A detailed account of the cooperation behavior of vampire bats is described in [9–11], but here we highlight the key findings. Vampire bats are fed by blood from mammals, they have no other food option. Vampire bats live in roosts but normally hunt alone. The success rate in finding food by themselves is relatively low and consequently it is not always guaranteed that there will be enough food to fulfill their needs. Thus, the vampire bats use cooperation in order to ensure they will get their daily food. All vampire bats that find food will store as much as possible and share later on with other individuals of their species. The other vampire bats are not necessarily family members but rather from a cooperation cluster the bats have exchanged blood with previously. Therefore, the first rule of the vampire bat is reciprocity – if you feed me now, I feed you later. The concept of reciprocity is described in detail in [12]. As cheating erodes cooperation vampire bats will exclude any bat from the cooperation cluster if cheating is detected. A bat that did not find any blood before returning back to the roost will request other vampire bats from the own cooperation cluster to feed it. In case this is denied, there are two options: the requested vampire bat either has not found food or it does not want to share its food with the requesting bat and rather use the blood found to extend his own cooperation cluster. The requesting vampire bat will therefore check with its nose in the stomach of the requested bat whether there is blood, and if blood is indeed detected, it will label this bat as cheater and never again exchange food with this individual. Vampire bats have very sensitive sensorial systems and relatively developed brains that allows identifying and remembering those cheaters for a lifetime. Therefore, the second rule of the vampire bats is the punishment of cheaters to keep cooperation ongoing. If free–rider bats would get their food for free, the cooperation among the vampire bats might fully break down. Punishment to promote cooperation is also known by humans [13]. A large number of experiments have been carried out showing that punishment in order to maintain cooperation is also performed at the cost of the own potential gain.

7.5 Monkeys

Cooperation between monkeys adds interesting insights to this discussion on cooperation. F. de Waal and his team have shown that monkeys invest into cooperation, as shown in [14, 15]. In one of the experiments two monkeys in a cage cooperated in order to get food. The food was placed on a heavy bar that only both monkeys together could pull towards the cage. But in fact the experiment enabled only one of the cooperating monkeys to collect the food as they were separated by a fence. Nevertheless, it was shown that the monkey getting the food shared it through the fence with the other monkey. The impressive outcome of the research was that monkeys often share more than half of the gained food with the other monkey even if the work they have jointly conducted to get the food was the same. In similar experiments with humans, the gain for a cooperating human was often less than the half. Therefore, monkeys seem to invest in future cooperation activities. Another fact the research of F. de Waal has shown is that the cooperation among monkeys depends heavily on whether monkeys are related or not. This finding is closely underlining the results of Hamilton [16] (see Chapter 8 for more details). Therefore, we can expect that cooperation will be easily established where cognition among cooperating partners exist.

7.6 Prisoner's Dilemma

In order to further understand the cooperation rules in Nature, the prisoner's dilemma is a great tool to explain the consequences of being selfish or cooperative. In 1950 M. Flood and M. Dresher introduced a game considering two prisoners interviewed by the police at the same time but without each other knowing their responses. Tucker developed the game further,

Table 7.1 Original Prisoner's Dilemma as proposed by M. Flood and M. Dresher.

		Prisoner 2	
		cooperates (stays silent)	defects (betrays)
Prisoner 1	cooperates (stays silent)	1 year prison for both	3 years for Prisoner 1 and Prisoner 2 goes free
	defects (betrays)	3 years for Prisoner 2 and Prisoner 1 goes free	2 years prison for both

The prisoner's dilemma for two players with four different possible outcomes. The Nash equilibrium is reached if both players defect.

specifying the rewards and penalties. Both prisoners have only to choose between two options, namely *defection* or *cooperation*. Based on the two simultaneous decisions the prisoners receive their penalty as given in Table 7.1. Defection means that one prisoner confesses to the police, trying to reduce the own punishment. By betraying the other prisoner the 3 years' punishment is avoided. Based on the decision of the other befriended prisoner, the time in jail will be limited to 2 years. In the best case the prisoner will be free at the price that the other prisoner is going to stay in for 3 years paying for his loyalty. The dilemma is in fact that the two prisoners cannot coordinate their actions and have always to fear the betrayal of the befriended other. Based on the Nash equilibrium [17, 18] both prisoners should exactly do that and betray the other prisoner, in order to get an acceptable punishment and avoid the highest punishment at the same time.

In 1980's Robert Axelrod [19–21] organized two tournaments around the prisoner's dilemma asking for computer program contributions that would play against each other to find the optimal strategy in case the game is repeated for several rounds instead of one round in the original prisoner's dilemma. This is referred to as iterative prisoner's dilemma. For the tournament Axelrod defined the reward points as given in Table 7.2. If both players cooperate the reward (R) is 3 points for both players. If one player defects and the other one cooperates the temptation (T) gave 5 points while the sucker's payoff (S) is zero. If both players defect both parties will receive at least 1 point only as punishment (P). The points are designed in the way that

$$T > R > P > S \tag{7.1}$$

Table 7.2 Prisoner's Dilemma proposed by Axelrod.

		Player 2	
		cooperates	defects
Player 1	cooperates	$R_1 = 3$ and $R_2 = 3$	$S_1 = 0$ and $T_2 = 5$
	defects	$T_1 = 5$ and $S_2 = 0$	$P_1 = 1$ and $P_2 = 1$

This is the prisoner's dilemma by Axelrod which was used for two of his tournaments.

and

$$2R > T + S. \tag{7.2}$$

The latter is needed as Axelrod wanted to play the iterative prisoner's dilemma and alternating strategies should not be seen as a potential candidate for cooperation. If Equation (7.2) would not hold, then the players could agree on letting each other win every second time, thereby achieving more points than by being cooperative all the time. In the first tournament Axelrod received several contributions that differed in their complexity. The simplest program had four lines of code while others had a few hundred lines of code. In the end the winner was a strategy called *tit–for–tat*. The strategy was rather simple and with four lines of code the less complex among all participating code examples. It behaves nicely in the very first round (cooperate) and then always repeats whatever the other player has played in the round before (tit–for–tat). This strategy was quite successful. Even in a second tournament, some years later, when tit–for–tat was already known to the research community, no strategy was found that could get better results than tit–for–tat.

Only twenty years later a research team from University of Southampton, UK were able to beat the strategy. As explained in [22] their approach was to send in multiple candidates. The clue was that the submitted candidates could recognize each other and one out of the candidates had a special task while the others behaved all the same. The special candidate was playing always tit–for–tat when detecting the foreign strategies, and defection when he/she recognized his/her own strategies. The other candidates of Southampton played cooperation when they recognized the special candidate but played defection otherwise. The main idea is to destroy the gain for the other strategies while optimizing the gain of the special candidate. This is a form of altruism as all candidates, beside the special candidate which won the competition, performed badly. Even tough tit–for–tat performed better than the mean value of all candidates by Southampton, it could not win over the special candidate. The example shows the importance of cognition in cooperative networks as the strategies had to recognize each other in order to collaboratively outperform tit-for–tat.

In order to give a first impression why tit–for–tat is successful we look into three different strategies and investigate how they would perform if played against each other:

cooperation In this strategy the player will always cooperate, hoping the others do the same. The players are agnostic to the outcome of prior games.

defection In this strategy the player will always defect, hoping for the largest reward. Also here the players do not change their strategy.

tit–for–tat In this strategy the player starts with cooperation and then repeats the last movement of the opponent. This strategy is adaptive and depends on the outcome of the games played before.

The rewards for two players playing against each other are giving in Table 7.3. These results are achieved if the game takes infinite steps. The points for the game between tit–for–tat and

Table 7.3 Reward points for the three different approaches of the prisoner's dilemma used by Axelrod.

	cooperation	defection	tit–for–tat
cooperation	3	0	3
defection	5	1	1
tit–for–tat	3	1	3

Here the mean points for three different strategies playing against each other are given.

defection depend on the number of iterations. The correct points for the defection strategy (against tit–for–tat) is $\frac{4+i}{i}$ and for the tit–for–tat (against defection) $\frac{i-1}{i}$. For large values of i, both values will converge to 1 as given in Table 7.3. From Table 7.3 we also see that the cooperation strategy is being exploited by the defection strategy. The game between defection and cooperation is giving the defection players always five points and zero for the cooperating players. Therefore, *credulous* cooperation is not the right strategy. We also see in Table 7.3 that the tit–for–tat strategy is robust against the defection guys even if it needs to adapt in the beginning. On the longer run both strategies will tie with one point each. Therefore, the outcome of the game depends on the number of iterations (which was high in the tournament).

Furthermore the results will change if the game is played with multiple players (more than two). The expected reward for the individual player depending on the own strategy and the number of players per strategy in the game is E_c (given by Equation 7.3), E_D (given by Equation 7.4), and E_{TFT} (given by Equation 7.5). The number of players performing the cooperative strategy is denoted as C, D for the number of defection players, and T for the number of players playing tit–for–tat.

$$E_C = 3 \cdot \frac{C-1}{T+D+C-1} + 3 \cdot \frac{T}{T+D+C-1} \tag{7.3}$$

$$E_D = 5 \cdot \frac{C}{T+D+C-1} + 1 \cdot \frac{D-1}{T+D+C-1} + 1 \cdot \frac{T}{T+D+C-1} \tag{7.4}$$

$$E_{TFT} = 3 \cdot \frac{C}{T+D+C-1} + 1 \cdot \frac{D}{T+D+C-1} + 3 \cdot \frac{T-1}{T+D+C-1} \tag{7.5}$$

We already mentioned the importance of the number of games played for the potential outcome of the game. The Equations 7.3, 7.4, and 7.5 give the points after a sufficient number of games. In order to investigate the initial phase of the games, we switch to simulations.

In Figure 7.1 the achieved points for two different scenarios are shown over the number of simulation steps. We use the NetLogo simulation tool [23] to test out two scenarios. The source code for the described PD scenario for the NetLogo tool is available at [24]. NetLogo is an agent-based simulation tool. Each agent has a given strategy and randomly meets other agents to play a game with. In the simulation the number of agents for a given strategy can be freely defined. In the first scenario ten players use the defection strategy while ten players use tit–for–tat ($T = 10$, $D = 10$, and $C = 0$). In the second scenario the number of players

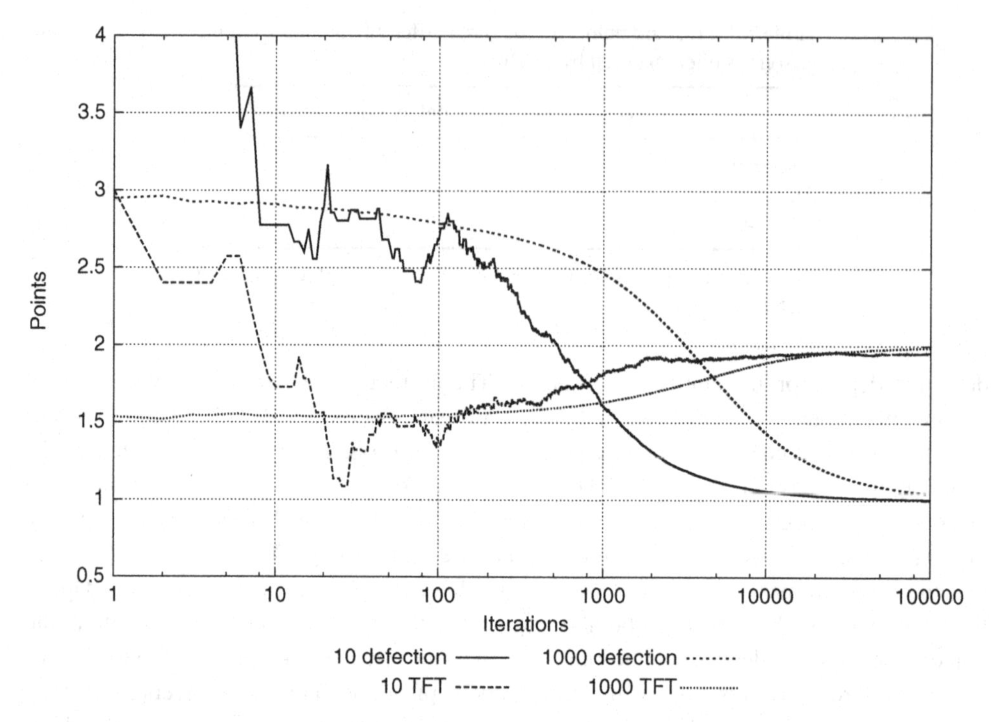

Figure 7.1 Points achieved by the prisoner's dilemma for the strategies *defection* and *tit–for–tat* for two different scenario settings.

is increased to one thousand for both strategies ($T = 1000$, $D = 1000$, and $C = 0$). As we mentioned beforehand the points depend on the iteration steps.

Figure 7.1 shows that defection always achieves better results for small number of iterations in both scenarios. The reason for that is that the players have to learn each other and change (in the case of tit–for–tat) their behavior. In the first scenario a few hundred simulation steps are needed until the tit–for–tat strategy outperforms the defection strategy. In the second scenario with one thousand players per strategy the number of simulation steps outperforming defection is increased to a few thousand. The point in time when the tit–for–tat strategy is achieving more points than the defection strategy is very important. The players need to learn about the other players' strategy. In prior work we referred to this as cognition. Furthermore, we see that the overall points for the strategies converge to one for the defection strategy and towards two for tit–for–tat. Owing to the size of the virtual world, the mobility of the agents and the numbers of agents, per simulation step the number of games for 20 agents is 1.37 and for 2000 agents 1807.

Equations 7.6 and 7.7 give the expected points $E(i)$ at any given iteration step i for the different strategies for any number of players D playing defection and any number of players T playing tit–for–tat. For large i the last fraction in each equation becomes one, but for the initial phase this term has still a significant impact. For the second scenario with 2000 nodes the points for the defection and tit–for–tat after the first iteration are 3 and 1, respectively.

With a larger number of iterations the points are 1.5 for defection and 2 for tit–for–tat.

$$E_{TFT}(i) = 3 \cdot \frac{T-1}{T+D-1} + \frac{D}{T+D-1} \cdot \frac{i-1}{i} \tag{7.6}$$

$$E_D(i) = \frac{D-1}{T+D-1} + \frac{T}{T+D-1} \cdot \frac{4+i}{i} \tag{7.7}$$

The prisoner's dilemma has implications for cooperation in a mobile cloud. As also mentioned in [21] by Hamilton and Axelrod, the probability to meet again has a severe impact on the expected reward. In the mobile cloud context this probability is not necessarily limited to the probability to actually meeting each other again. In such a case this depends on the reciprocal cooperation activity. E.g., in case of cooperative download (as given in Chapter 9.2) by two devices, each device downloads one non-overlapping part of the target content via the cellular link and exchanges it with its counterpart device on the short–range link. The exchange on the short–range link is realized by a large number of IP packets. In case the other peer is not cooperative anymore it is easy to stop the cooperation. In this case no harm has been done as the cooperation is established on a per packet basis. The investment is small as is the gain for a cheater. Things change if we consider mobile hot-spot scenarios. Here one user could be for example downloading a large file via a cooperating device. Based on reciprocity, the user being helped offers an equivalent service to the helping user in the future. Here the meeting probability is not on a per packet basis but more on a per session basis. The likelihood to get exploited is larger than for the cooperative download case. In the mobile cloud context there is even another angle we have to look at. Owing to the error-prone wireless link, defection might be detected where actually just errors on the wireless link occur. Therefore, our strategies need to be more robust. In [25] Axelrod discussed the performance of the iterated prisoner's dilemma in presence of noise. This is very important for cooperation in wireless networks that are prone to packet erasures. Such erasures could lead to false detection of defection of cooperation and destroy the performance. Therefore, the algorithms need to be robust and less strict knowing about the nature of the wireless medium.

7.7 Conclusion

In this chapter we have discussed the basic rules of cooperation by investigating the behavior of hyenas, cheetahs, vampire bats, monkeys and others. The derived rules are quite simple and readily applicable to mobile devices and communication protocols. Furthermore we have introduced the basic concepts of the prisoner's dilemma, investigating different strategies. One basic rule of thumb is that cooperation will flourish whenever reciprocity at high rate is achievable.

References

[1] O. Leimar and A.H. Axen. Strategic Behavior in an Interspecific Mutualism: Interactions between Lycaenid Larvae and Ants. *Animal Behavior*, 46:1177–1182, 1993.

[2] S.J. O'Brien, D.E. Wildt and M. Bush. The Cheetah in Genetic Peril. *Scientific American*, 254(5):6876, 1986.
[3] Michigan State University. Now take a Look at what happens when Spotted Hyenas cooperate against a Group of Lions. http://museum.msu.edu/exhibitions/Virtual/hyena_kiosk/HyenaCooperation. html.
[4] Duke University. Hyenas cooperate, problem-solve better than Primates. http://today.duke.edu/2009/09/hyenas.html, 2009.
[5] C.R. Darwin. The Origin of Species. *The Harvard Classics. New York: P.F. Collier & Son*, XI:1909–1914, 2001.
[6] R.L. Pitman and J.W. Durban. Cooperative Hunting Behavior, Prey Selectivity and Prey Handling by Pack Ice Killer Whales (Orcinus orca), Type B, in Antarctic Peninsula Waters. *Marine Mammal Science*, 28(1):16–36, January 2012. http://dx.doi.org/10.1111/j.1748-7692.2010.00453.x.
[7] D. Wiley, C. Ware, A. Bocconecelli and D. Cholewiak. Underwater Components of Humpback Whale Bubble–net Feeding Behavior. *Behaviour*, 148(5):575–602, 2011.
[8] T. SimilÃd' and F. Ugarte. Surface and Underwater Observations of cooperatively feeding Killer Whales in Northern Norway. *Canadian Journal of Zoology*, 71(8):1494–1499, August 1993.
[9] F.H.P. Fitzek and M. Katz, editors. *Cooperation in Wireless Networks: Principles and Applications – Real Egoistic Behavior is to Cooperate!* ISBN 1-4020-4710-X. Springer, April 2006.
[10] M. Ridley. *The Origins of Virtue : Human Instincts and the Evolution of Cooperation*, 1998.
[11] G. Wilkinson. Reciprocal Food Sharing in Vampire Bats. *Nature*, 308:181–184, 1984.
[12] V. Smith, E. Hoffman and K. McCabe. Reciprocity: The Behavioral Foundations of Socio-economic Games. *Springer-Verlag – Understanding Strategic Interaction - Essays in Honor of Reinhard Selten*, pages 328–344, 1997.
[13] N. Raihani, A. Thornton and R. Bshary. Punishment and Cooperation in Nature. *Trends in ecology & evolution (Personal edition)*, 27(5):288–295, May 2012.
[14] S.F. Brosnan and F. de Waal. Monkeys reject Unequal Pay. *Nature*, 425:297–299, 2003.
[15] S.F. Brosnan, H.C. Schiff and F. de Waal. Tolerance for Inequity may increase with Social Closeness in Chimpanzees. *Proceedings of the Royal Society, B-Biological Sciences*, 272:253–258, 2004.
[16] W.D. Hamilton. The Evolution of Altruistic Behavior. *The American Naturalist*, 97:354–356, 1963.
[17] C.A. Holt and A.E. Roth. The Nash Equilibrium: A Perspective. *Proceedings of the National Academy of Sciences of the United States of America (PNAS)*, 101:3999–4002, 2004.
[18] J.F. Nash. Equilibrium Points in n–Person Games. *Proceedings of the National Academy of Science of the United States of America (PNAS)*, 36:48–49, 1950.
[19] R. Axelrod. *The Evolution of Cooperation*. basic Books, 1984.
[20] R. Axelrod. *The Complexity of Cooperation*. Princeton Paperback, 1997.
[21] R. Axelrod and W.D. Hamilton. The Evolution of Cooperation. *Science*, 211:1390–1396, 1981.
[22] A. Rogers, R.K. Dash, S.D. Ramchurn, P. Vytelingum and N.R. Jennings. Coordinating Team Players within a Noisy Iterated Prisoner's Dilemma Tournament. *Theoretical Computer Science*, 377(1-3):243–259, May 2007.
[23] U. Wilensky. NetLogo pd n–Person Iterated Model. Center for Connected Learning and Computer-Based Modeling, Northwestern University, Evanston, IL, 2013.
[24] Frank Fitzek. Pd n–person iterated. www.fitzek.net, May 2013. NetLogo source code. Edits made by Fitzek to the original PD N–Person Iterated version by Wilensky.
[25] J. Wu and R. Axelrod. How to cope with Noise in the Iterated Prisoner's Dilemma. *Journal of Conflict Resolution*, 39:183–189, 1995.

8

Social Mobile Clouds

Real egoistic behavior is to cooperate!

Kurt Edwin

This chapter introduces the social component of the mobile clouds, exploring different ways to foster cooperation among selfish users. Different forms of cooperation are introduced and compared with each other. One of the key elements of this chapter is the introduction of social rewards for being cooperative. This chapter also discusses the evolutive development of social and mobile networks, highlighting in particular how these networks are becoming increasingly dependent on each other.

8.1 Introduction

Mobile clouds are dependent on the willingness of mobile users to cooperate. Without cooperation there is no way to build mobile clouds and each user would be independently acting on his own. User cooperation is the underlying principle of mobile clouds. Therefore it is critical to understand the reasoning behind cooperation to ensure users' willingness to participate. This helps to develop effective cooperative strategies and to create attractive situations motivating users to cooperate. In general, there are fundamental differences in the design and operation of mobile clouds, depending on whether users are forced to cooperate, or they are willing to invest into cooperation for whatever reason, or if the users see a clear benefit in cooperating with others. Cooperation rules in Nature, discussed in Chapter 7, are helpful, but here we investigate potential cooperation cases and how we can stimulate them. Following the quote of this chapter *Real egoistic behavior is to cooperate!* one can develop cooperative strategies for mobile clouds such that the gain for each participating user is clear. This is one of the essential characteristics of mobile clouds, that is, they are in principle designed to provide each and

Mobile Clouds: Exploiting Distributed Resources in Wireless, Mobile and Social Networks, First Edition.
Frank H.P. Fitzek and Marcos D. Katz.

every user with a certain benefit, impossible to attain while operating in an autonomic (i.e., non–cooperative) manner. However, there are also cases where there is no involved gain at all or the gain is not easily visible to the user. The benefits achieved by cooperating in a mobile cloud can take place in different domains, as discussed later in this chapter.

8.2 Different Forms of Cooperation

In current communication systems users themselves have limited connectivity freedom, they can at the most decide whether to be connected to the cellular or WLAN network, when both are available. Certainly, a user cannot choose a particular cellular base station to be connected to. In [1] we compared this situation to slavery as the mobile devices, even though they might be called smart, they are dumb end devices.

In mobile cloud–oriented communication systems this situation will change. Regardless of the fact that cooperation among users is likely to be beneficial from a system perspective, mobile users might still choose to decline cooperation as their own investments into the cooperation could be larger than the benefits. This is referred to as selfish behavior and it is totally fine as it protects the user from getting exploited.

From the system standpoint there is a huge potential to improve the performance in a mobile cloud, if we would have full control over the mobile devices. This has been already shown in [2–4] for data throughput, energy and delay. It is interesting to notice that even in a worst-case situation, the performance of devices in a properly designed mobile cloud should never be worse than in current state–of–the–art (non–cooperative) solutions. But these are just toy examples, and in general the system performance should improve significantly whenever cooperating devices join forces. Consequently, one might well argue that devices should be forced into cooperation without involving the user. But this is not necessarily in line with the current trend seen in the market of cooperative services. In fact, today new services for mobile devices enter a mobile platform via the app marketplaces. Therefore, it is ultimately the user who is in charge of installing the app that makes the device a cooperative device. The user also decides whether to activate this app based on one and only one criterion, the so–called cost–benefit balance.

The cost–benefit balance is a simple, though important decision weighing the involved costs for undergoing cooperation on one hand, and the expected benefit on the other hand. For instance, in a cooperative relaying scenario, a forwarding node would invest its own bandwidth share and energy to help other nodes without obtaining any gain in return. So, one may question the reason for this individual node in using some of its resources for doing the relaying job. A great deal of research has been carried out in the field of cooperative relaying assuming that the relaying node will do the job just because it behaves *nicely* and believes in the system gain. Certainly, this can be seen as a case of forced cooperation, and a typical example would be the use of dedicated nodes built for that purpose owned by the same user. Network operators employ such repeaters (nodes forced to cooperate by their owners) to increase coverage or enhance cell edge performance. The cost–benefit balance is evaluated by each node separately and therefore for some nodes cooperation is a valid solution while others might decide to skip

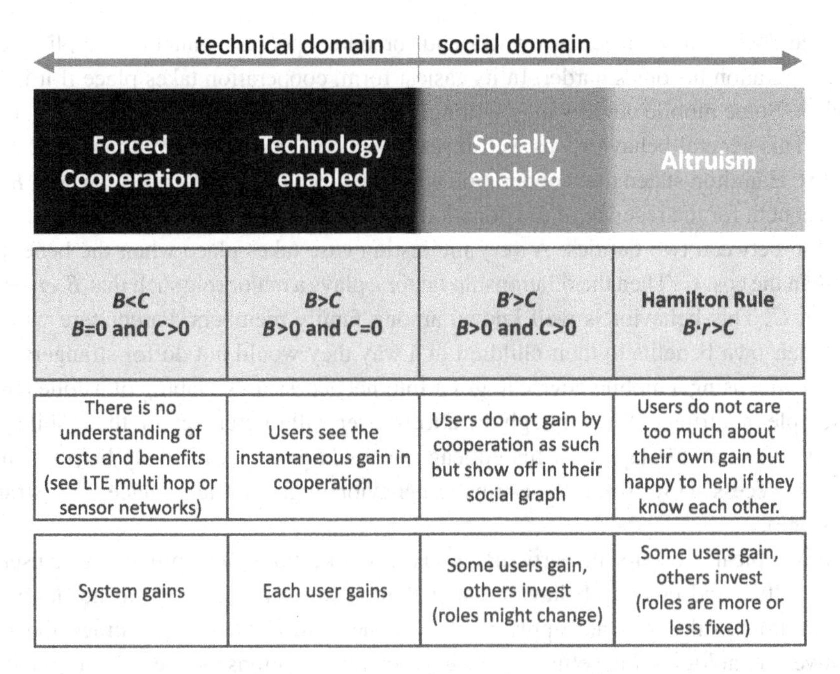

Figure 8.1 Different forms of cooperation.

cooperation. In this chapter we first look into different forms of cooperation, comparing them with each other. Then we derive some methodologies to foster cooperation by adjusting the cost benefit–balance.

Figure 8.1 defines four different forms of cooperation, namely forced cooperation, technology–enabled cooperation, socially–enabled cooperation and altruism. In this chapter we limit ourselves to consider these four forms as they represent the dominating cooperation forms. In general, and regardless of its type, cooperation has a cost (C) and a benefit (B) that the participants are paying or receiving, respectively.

A short example should help the reader to understand this relationship. The costs of cooperation might be extra energy that is invested to power up a second air interface in order to communicate with other devices. The benefit, on the other side, might be for the case of cooperative download, an energy saving on the overlay air interface as less data has to be retrieved from there. The relationship of these two parameters, namely C and B, as mentioned before, affects the incentive and ultimately establishment of cooperation.

Forced cooperation describes the situation where nodes have to follow system rules rather than making individual decisions. Forced cooperation takes place for instance in the case of LTE relay networks, or when the owner of multiple devices forces his or her devices to cooperate in order to get the best system performance rather than considering the individual device, e.g., sensor networks. Here the cost for the individual device is large but the benefit is very small or not present at all, therefore $C > B$ as $B = 0$. Nevertheless, the mobile devices or nodes have no choice as they are forced into cooperating, in other words slavery.

If mobile devices have different owners and forced cooperation cannot be applied, establishing cooperation becomes harder. In its easiest form, cooperation takes place if it is based on altruism. Some mobile devices may willingly sacrifice some of their own benefits in favor of others. This general behavior was well investigated and modeled by William D. Hamilton in 1963 [5]. Hamilton stated that cooperation will take place in case $B \cdot r > C$, where B is the expected benefit for the receiver of the donation, C the cost involved for the donor and r is the relationship between two entities. A very interesting case takes place when the benefit B is smaller than the cost C. Then the relationship factor r plays a major role such that $B \cdot r$ become larger than C. This behavior is well known among family members. Parents are willing to sacrifice their own benefits to their children in a way they would not do for strangers. In the mobile domain, using a mobile device to grant Internet access to the laptop of a colleague is a great example of altruism. Even though the altruist user will lose energy on the mobile phone to support access on the laptop of his colleague (r is large), he will likely help by sharing his wireless access. However, such altruistic behavior might not take place with strangers (r equals one).

Altruism typically occurs in particular scenarios, like home and office. Most users are driven by selfish and egoistic behavior rather than altruism if the users do not know each other. Hamilton's rule does not apply if $r = 0$. The easiest way to encourage the use of cooperative technologies for selfish users is to create situations where the instant benefit (B) is larger than the cost (C) of for all participating users. Cooperation would happen automatically as *Real egoistic behavior is to cooperate!*. Such a situation can be created by using the proper technology and is therefore referred to as technology–enabled cooperation. Cooperative video streaming as [6] exemplifies such a scenario. For each additional participant the energy consumption per mobile device can be reduced while the quality–of–service may be maintained or eventually increased. Nevertheless, there are situations where the local exchange consumes significant resources. For instance, if the local cluster is connected via multi hops instead of single hops, the overall energy consumption for a single device of the mobile cloud could be larger than for a standalone device. Such a situation may occur when more than two entities cooperate. The concept of network coding, discussed in Chapter 5, can be used to reduce the cost for the local cooperation, enabling us to use cooperation in a wide number of situations and scenarios.

Should the benefit of cooperation still remain unclear to some users, the concept of *social reinforcement* should be applied. Not much research has been carried out on this kind of cooperation. In such a collaborative scenario, we predict that one or more entities gain from the cooperation (we call them the beneficiaries) while one or more entities invest into cooperation but do not gain anything out of it (we call them the donors). This concept is similar to Hamilton's altruistic approach, but here we assume the players do not know each other and therefore the parameter r equals one. While the gain for the beneficiaries ($B > C$) is clear, the gain for the donors is not ($B < C$). It is a fact that most users of smartphones are members of social networks such as Facebook. When donors help establishing cooperation with little–to–no perceived benefit for them, their efforts could instead be rewarded in a different domain, namely in social networks with a suitable type of benefit (B). This can be done by simple notification or other gamification concepts as we lay out later in this chapter. The cooperative involvement of donors, measured with numbers, plots or icons, could be made readily visible in donors' social networks. Such an open rewarding system could serve to further motivating users to be cooperative, as such an attitude will be noticed by their social entourage. Therefore, the gain for the donors is within the social domain where they can obtain rewards from the receivers as well as their existing social graph.

8.3 Social Networks and Mobile Clouds

Mobile communications systems have undergone a tremendous evolution in the last decades as discussed in past chapters. Originally, the main service provided by mobile communication systems was voice. However, the paradigm shift from analog to digital systems in the second generation of mobile communication systems introduced security and mobility services and, more importantly, it opened the door to providing data services. With the introduction of the third generation (3G), data services have been identified as a key application field, although it was originally envisioned to support mobile web browsing with data rates limited to a few megabits per second. In part motivated by the need to support data-hungry services in the mobile world, the fourth generation (4G) of mobile communication systems currently offers higher data rates, up to hundreds of megabits per second. Important from our perspective is the support of services allowing interaction and sharing of rich content in social networks.

Social networks started in the Internet in the middle of the 1990's, e.g., classmates.com in 1995, but it was the boom of Web 2.0 in the early 2000's that finally allowed social networks to become a big player of the Internet, e.g., myspace.com in 2003. In 2004, facebook.com was introduced, becoming a game changer in social networks. While the first social networks only targeted people's basic connectivity needs, e.g., to stay in contact, the new generation of social networks became an essential part of the daily life of the users. The scale of this day–to–day interaction is enormous and evidenced by the fact that Facebook has over one billion users nowadays, being one of the top two largest traffic generators globally in the Internet according to Alexa (alexa.com).

Although social and mobile networks have become embedded in our daily lives, they are the offspring of two very different processes. While mobile networks are highly standardized, planned and regulated, social networks follow a perhaps more chaotic and anarchic approach, epitomized by Facebook's philosophy to *move fast and break things* to understand and adapt quickly. Nowadays the evolution path of these two networks has become more and more harmonized and intertwined, and this converging trend is expected to continue in the future. Indeed, more than ever social networks work and ultimately depend on wireless and mobile communications technologies. The opposite is also true, more and more communications networks, mobile devices and services are designed to support rich social network interactions.

The evolution of the social and mobile networks can be understood by dividing it into three main phases, as given in Figure 8.2. The key to this division is to recognize these two networks started operating separately, then started to coexist and have initial interactions, and in the future they will stimulate each other providing revolutionary data communication designs and novel possibilities for higher interactions in social networking that may go beyond the digital data-sharing domain.

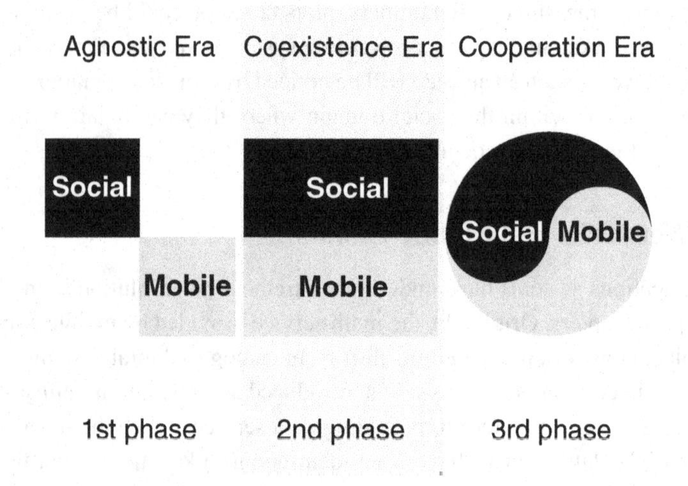

Figure 8.2 The three main phases in the evolution of social and mobile networks.

8.3.1 Evolution Phase I: Agnostic Era

In the beginning, only browser-based social networks were available and users entered their social networks mainly with their PCs or laptops. Mobile and social networks were initially following different developing tracks due to a number of factors. First, the lack of Web 2.0 capabilities in mobile web browsers. Second, mobile phones supported low data rates, had small display sizes and low processing power compared with current mobile device technology. Third, back then social networks were less ubiquitous nor popular in our daily life, and they were not designed to create on the user a need or desire to be connected or just check on the go. The underlying assumption then was that the rate a user frequented the social network was low as a PC or laptop would be required to gain access. Social networks were in the beginning only popular among certain groups, and to a great extent social interaction was initially carried out using other media than social networks, like email, personal web pages and phone calls. This stage, characterized by virtually no practical interplay between mobile and social networks, is illustrated on the left-hand side of Figure 8.2. Networks had significantly different use cases at that stage.

8.3.2 Evolution Phase II: Mobile Networks Supporting Social Networks

With the introduction of Internet–enabled mobile devices, users of social networks did not only want to consume the content of the social networks on the move, but also they realized that the mobile device itself was a great tool to produce and contribute valuable content to the social networks. In this phase, it was technically possible to use social networks on top of mobile communication systems, as given in the middle of Figure 8.2.

The very first social network that understood the power of mobile devices for social networking was foursquare.com, an evolution from a previous service (Dodgeball), which relied on Short Message Service (SMS) technology [7]. foursquare.com provided a service that enabled users just to share their current location. At that time, mobile devices already supported location retrieval by GPS, which made it very simple for the users to share their location with friends. foursquare.com was not the first that tried this service, but it added some gamification into the social network, allowing users to collect points by checking in at certain locations. Users with the most check–ins at a given location could become the mayor. This simple incentive inherently created a desire to use it on a daily basis, which made the service very successful.

Moving to the mobile space was quite easy during this phase, because the mobile browser became better and supported more functionalities. But beyond this browser support, mobile devices provided additional native support, namely allowing the distribution of mobile applications (apps) by third parties. Of course, native support was by no means immediately widespread. For example, Apple initially believed that they could offer all services in the mobile browser, which meant that the very first iPhone had no mobile apps. There had been a long-standing and still ongoing battle between mobile browsers and mobile apps to decide which is the best service-providing platform. For example, Facebook remained in the mobile browser domain, neglecting the mobile app space for quite some time.

The advantage of the web-centric social network was that it could run in any browser and therefore the coverage was not limited. However, native applications had more access to mobile device resources, e.g., location, camera, some of which are essential parts for enabling features of most social networks. Moreover, an added advantage is the ability to work off–line, something that browsers could not do at the time. Interestingly, current efforts on introducing HTML5 should allow the browser to get more access to native resources and work off–line, amongst others. This would allow the browser to compete with native applications.

Shortly after the mobile devices opened up and developers were able to make their own programs through the mobile device application programming interfaces (APIs), even the social networks started to offer their own third-party APIs. This paved the way for a real interplay of social and mobile networks.

Learning from the mistake of sticking too long with the browser, Facebook is currently trying to link the social network even deeper into the mobile domain, introducing *Facebook Login* and *Facebook Home*. The first aims at replacing any mobile app specific login with a general login procedure based on their social network infrastructure. Of course, Facebook is not the only social network implementing this approach. On the other hand, *Facebook Home* is an even deeper integration of the social network into the mobile network by providing content directly without opening a browser or an application.

One of the downsides of social networks to mobile networks lies in the sheer amount of content generated by its users, having a particular large impact on network traffic. Even though new 4G technologies are advertised to provide data rates as high as 100Mbps, those figures are still shared (or aggregated) data rates amongst the different users in the cell. An individual user can enjoy such high data rates only if the cell is slightly loaded (e.g., few connected users). Users tend to post on their social networks whenever they are in crowded environments, just to share in real–time their experiences, such as following a sport event in a stadium. Vodafone reported recently that the traffic at the Oktoberfest in Munich doubles every year and the trend is not only to post simple pictures but to send entire videos. This becomes problematic as there could be tens of thousands of visitors across very small areas.

The use of mobile devices can provide more than just data transport for the social networks' content. Data collected from mobile phones was identified as a key source to understanding relational dynamics of individuals as well as providing insight into a society's social network at different scales. For example, [8] examined the communication patterns of mobile phone users using a series of electronic databases, including phone and e–mail logs, to understand the local and global structure of their underlying social network. [9] showed that mobile phone patterns allow us to accurately infer 95% of friendships and validated these results using self–report survey data provided by the mobile phone users. More detailed inferences of people's activities and interactions have also been considered by the CenceMe application [10].

From a different perspective, bringing together information from online social networks into mobile smartphones opens up interesting possibilities for more complex interactions. Early applications such as Dodgeball, relying on SMS technology, provided meaningful data for analyzing social interactions and the effect of said applications on those interactions [7]. In a sense, the availability of systems combining social network knowledge and mobile technology can have a significant impact on the underlying social structure and social interactions. WhozThat [11] constitutes another example of a system that exploits an infrastructure that shares social networking IDs locally, using wireless technology. WhozThat exploits wireless connections to access online social networks, binding identity with location in order to know more about a particular person you encounter. The infrastructure also has the potential to scale to support more complex context–aware applications.

Beyond providing the social network with a better knowledge of its own structure and interactions or local, context–aware services, active users could share information in order to build novel, large–scale services. This is often referred to as crowd–sourced information. An example out of several such initiatives is the social network Waze (waze.com). Waze is a navigation app that allows users to route efficiently from point A to point B. Besides giving the route information, Waze also considers current traffic situations. In contrast to other solutions, the traffic information is not retrieved from a centralized service with fixed mounted sensors, but it relies on measurements done by the users of the social network. Each active user is both receiving up–to–date information and feeding back information about its location and current speed. This crowd-sourced information is filtered, evaluated, and fed back to the users, providing them with a lively view on the current traffic conditions. Crowd-sourcing will become a key enabling concept in the third phase of interactions between social and mobile networks. The concept of crowd sourcing is very powerful and applications of it are increasingly being developed. Very recently Google started offering crowd–sourced services in their latest version of Google Maps.

8.3.3 Evolution Phase III: Deep Integration: Interplay of Social and Mobile Networks

As suggested on the right-hand side of Figure 8.2, the future will bring a real deep interplay between the two networks. Social networks will be a main driver to build mobile networks, going beyond cellular–oriented infrastructure. To understand the third phase we need to explain the mobile network evolution path. So far mobile networks have been built using a centralized cellular structure, where each device connects in a point–to–point fashion, as discussed in

Chapter 2. That is, each mobile device needs to connect to a serving base station in order to set up a data connection.

With the introduction of Wi–Fi capabilities on the mobile devices, Internet connections through short–range networks are now possible. As network operators face more and more challenges to support the high data rates needed by the mobile devices, due in part to the high costs of providing 4G infrastructure, they consider using *Wi–Fi offloading* strategies, also known as femto–cells or small cells. Of course, there are typically a large number of additional Wi–Fi hotspots, especially in cities, though not all of them are accessible. Most people use Wi–Fi hotspot at home and at the office, but in any other location they tend to fall back on the cellular connection. This limited access to other's Wi–Fi hotspots is due to security and legal issues. In many countries, the owner of an access point is responsible for the traffic that is transported over it. Thus, even altruistic people are compelled to close down their access point. The issue is trust. But this is exactly what social networks can provide.

In order to illustrate this idea of social–enabled technology, we take the mobile hotspot as an example. We assume that one mobile user has a flat rate connection to Internet and that one user has a limited or costly connection to Internet. If those two users span a mobile cloud the user with the flat rate allows the neighboring device to use him as a relay. The relaying device shares its data rate, increasing the energy consumption for the relay. Therefore he should be rewarded in some other way in order to cooperate. In the example given in Figure 8.3, Anna has no Internet connection. Frank offers his Internet connection on 3G via Wi–Fi using a

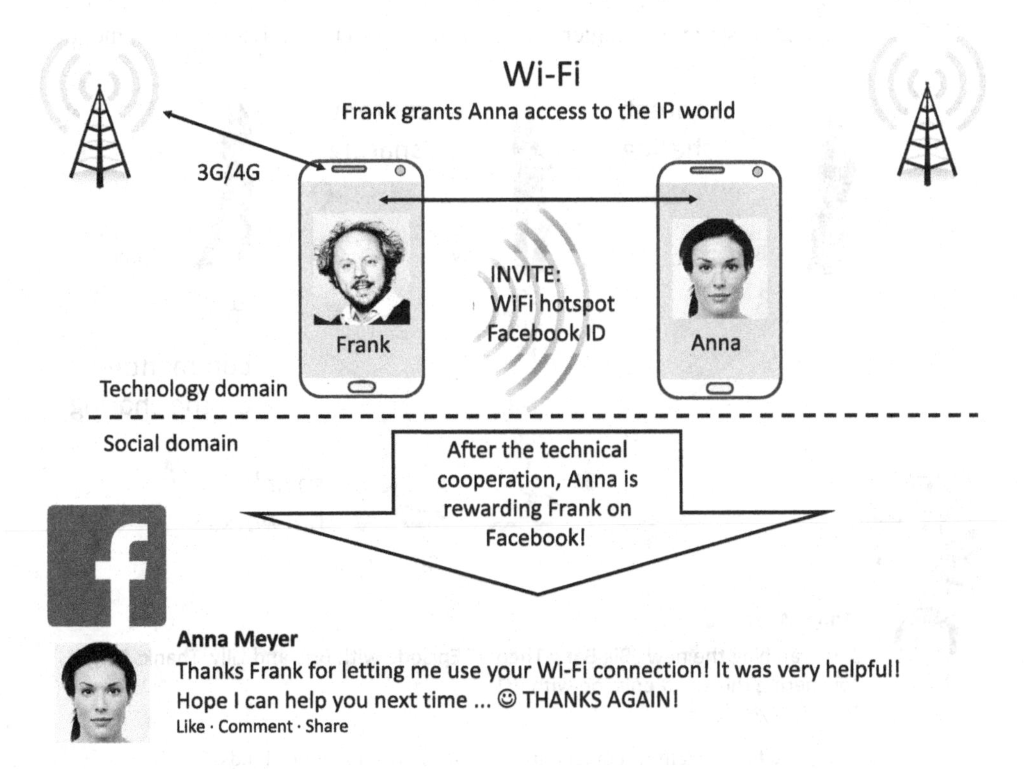

Figure 8.3 Impact of the social domain on mobile clouds.

special mobile app that is also in Anna's mobile device. Anna is made aware of this service by an invitation that Frank sends. In addition to the invitation to use his Internet connection, Frank also conveys his Facebook ID. During or after Anna has used Frank's Internet connection, she might want to thank him via Facebook by writing something on his wall. Frank is rewarded in the social domain for his cooperation in the technology domain as now Frank's whole social entourage knows about his cooperativeness. There is the possibility that Frank and Anna both have access to the Internet via 3G. Now there would be a direct link from Anna to her network operator. A Wi–Fi connection will still be established between Anna and Frank to timeshare their connections for optimal use of the cooperative communication channel. This would be an example of technology-enabled cooperation as, in this instance, both users gain. Nevertheless, using the social domain as a means of reward could also positively reinforce cooperative behavior in the future. The above example can be extended to consider rewarding given by operators. Indeed, if a user opens his/her mobile device for cooperation, the user is indirectly helping the operator to provide better service. Such behavior can be easily measured by the operator, and incentives proportional to the amount of helping provided can be readily provided.

While forced cooperation and altruism are two well–known concepts that represent the state of the art, technology–enabled and socially–enabled mobile clouds are undoubtedly beyond state of the art. In contrast to any other tit–for–tat cooperation scheme implemented in FON access points or BitTorrent, the cooperative exchange is not repaid with equal currency but is rewarded within a new dimension: the social domain.

A second example is given in Figure 8.4. This scenario was the ground idea for the project for NBC [12]. Instead of streaming content from an overlay to the individual users, the idea

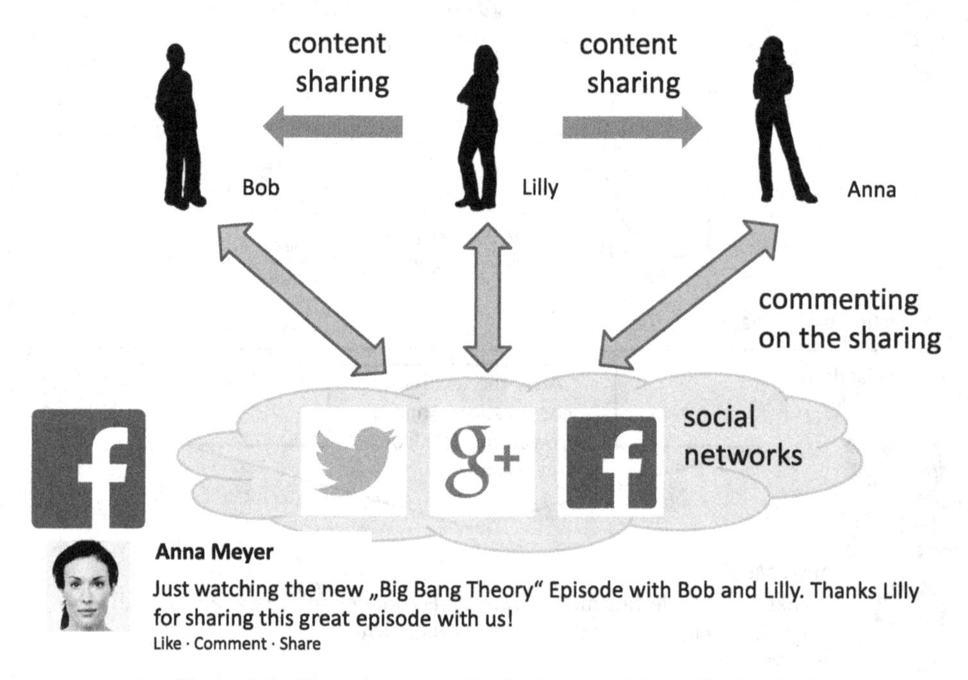

Figure 8.4 Example content distribution in social mobile clouds.

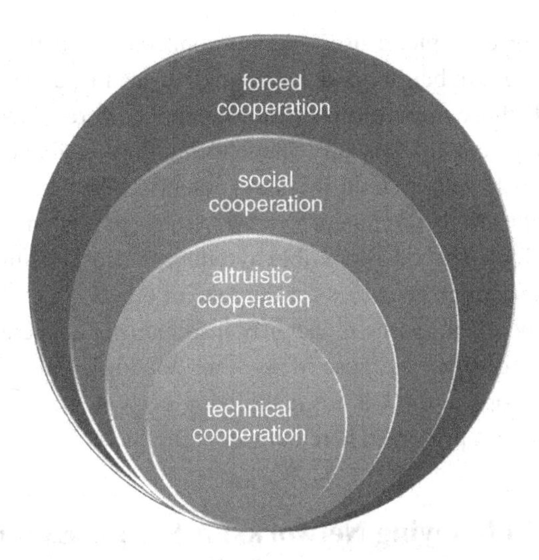

Figure 8.5 Different forms of cooperation.

was to enable users to disseminate the content among friends. In Figure 8.4 we see the user Lilly streaming an episode of e.g., *Big Bang Theory* to the close–by friends Bob and Anna. While they are watching or even after that they use the social networks to promote the content to the social graphs of all these three friends. From a content owner point of view, the local exchange has the advantage that their own servers are less used, which in turn results in less cost for the content owner. The monetization of such architecture is more complicated. Over the past years a number of attractive methods to monetize have appeared, offering new alternatives to digital rights management (DRM), such as advertisements, in app purchase, etc. For all these alternative ways, the method on how the content arrives to the users is of secondary importance. More important is the efficient dissemination of content, reaching a large number of consumers in a short time.

After introducing the four different forms of cooperation, we would like to put these approaches into perspective, as shown in Figure 8.5. As discussed, the easiest way to establish cooperation is to build technology in such a way that each cooperative participant gains. Technology–enabled cooperation is the most convincing form of cooperation. Each participant sees a benefit larger than the involved cost. For this kind of cooperation the relationship among cooperative users is not important as each participating user obtains a benefit. We claim that this sort of cooperation is the essential way of cooperation, as given in Figure 8.1. It also shows the need to consider new technologies in order to make this field of cooperation even more applicable. If technology–enabled cooperation does not apply, the next form of cooperation is altruistic cooperation. Here some users will gain and others have to invest into the cooperation. But as we pointed out, this only applies if there is a trusted relationship between nearby users (e.g., family members, friends, colleagues). The next level of cooperation is then socially–enabled cooperation, where a benefit in another domain should be found. The advantage of this approach is that it may also rely on persons that we know, and in addition, it applies to machines as well. For example, a user can get connected to a Wi–Fi access point in a café just by clicking the *I like* button "a la Facebook" instead of undergoing different ways of authorization [13]. If those three forms cannot be applied forced cooperation is the only way out. In the following we will focus only on the technology–enabled and socially–enabled cooperation as we still believe they hold the most potential. We emphasize again that these cooperation forms are not strictly separated. Even if we have to rely on social cooperation, we should make the technology as good as possible to reduce the cost and increase the likelihood that users are willing to sacrifice resources for the sake of others.

Intentionally we did not mention the monetarily inspired user cooperation (see [14]). With the introduction of ad hoc networks those ideas were floating around, but were never realized. Recently some startups in the US are revitalizing such ideas. Those initiatives are not cannibalizing the approaches presented here but extending them.

8.4 Cooperation in Relaying Networks: A Simple Example

For decades relaying techniques have been discussed at a system level without considering the individual users. Each node was forced to contribute to relaying for the sake of the system gain and with the promise that in the end each node will benefit. It was known already that some nodes have to invest into cooperation while others gain. But the assumption was that in the long run everybody will gain as the role of those who invest and those who are gaining will change over time.

A small example should underline the difference between different cooperation forms. We keep the example as simple as possible for the sake of clarity. The communication topology is composed of an access point that is connected to the Internet and allows mobile devices to connect to it using e.g., IEEE802.11 technology. As given in Figure 8.6 mobile device 1 is connected directly with the access point at a given data rate R_1. The second mobile device has either no connection to the access point or a connection at rate R_2 which is lower than R_1. The connection from mobile device 2 to mobile device 1 is labeled $R_{2\leftarrow1}$.

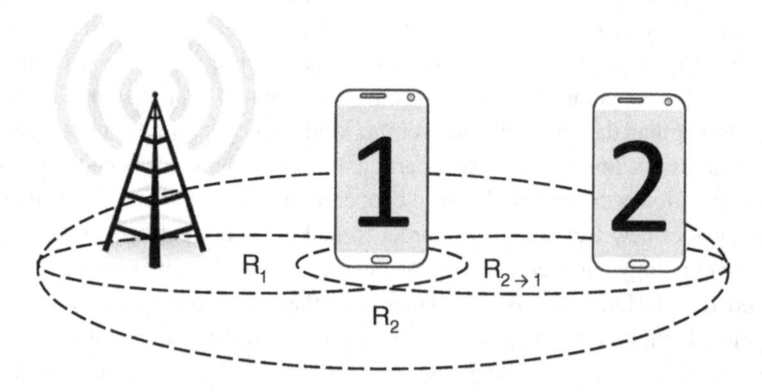

Figure 8.6 Relaying example showing different forms of cooperation.

First let us consider the case where mobile device 2 has no connection to the access point; the only way to convey information to the Internet is to relay its packets via mobile device 1. Mobile device 1 will only forward the packets if

1. mobile device 1 has no choice and is **forced** to do the relaying by the network operators' or manufacturers' settings,
2. mobile devices 1 and 2 know each other and due to Hamilton's rule [5] mobile device 1 is acting **altruistically**,
3. **social** elements such as Facebook, Twitter, etc. are used to build up trust among the mobile devices so that mobile device 1 gets interested in relaying,
4. or mobile device 1 believes in the system gain provided by the relaying **technology** and that it will gain on the long run despite the long pay–off delay.

Now we consider the case where mobile device 2 has a direct connection to the access point but this time relaying via mobile device 1 is beneficial for both devices. In an IEEE802.11 environment both devices will get the same channel capacity, in terms of transmission slots, despite the channel quality or data rate they are sending with. Owing to the lower rate R_2 compared with R_1, mobile device 2 will spend more time on the medium than mobile device 1. In case of $R_1 = 54$Mbps and $R_2 = 9$Mbps the time on the medium by mobile device 2 is six times larger that for mobile device 1. Or in other words, mobile device 1 occupies the channel only 14.28% of the time, while mobile device 2 is using 85.71%. If $R_{2\leftarrow 1}$ is significantly larger than R_2, mobile device 2 should send its packets to mobile device 1 for relaying. Mobile device 1 would forward the packets to free the channel for its own transmission. E.g., if $R_{2\leftarrow 1}$ equals 36Mbps, the system capacity would double. Mobile device 1 would double its time on the medium as it has to transmit its own packet and the packet from mobile device 2, both with rate R_1. The time to transmit the packet from mobile device 2 to device 1 takes 1.5–times longer. Thus the time to transmit two packets from two devices to the access point is reduced from $7 = 1 + 6$ to $3.5 = 1 + 1 + 1.5$. Therefore here we have a clear gain motivated by technology for instantaneous cooperation gains (without large pay–off delays). In other words, the connection of mobile device 2 gives the device leverage to *force* mobile device 1 to cooperate.

8.5 Conclusion

This chapter discussed the importance of social interactions between users in a mobile cloud. The social dimension offers new ways to enforce cooperation among users even without a real reciprocal gain. Cooperation will for sure not be only based on social elements, but it adds a very attractive angle to the set of cooperation enablers. The interested reader is referred to [15, 16] for further reading and implementation issues. Furthermore, the combination of the social domain with network coding, as introduced in Chapter 5, is further discussed in [17].

References

[1] F.H.P. Fitzek and M. Katz, editors. *Cognitive Wireless Networks: Concepts, Methodologies and Visions Inspiring the Age of Enlightenment of Wireless Communications*. ISBN 978-1-4020-5978-0. Springer, July 2007.

[2] G. Ertli, A. Paramanathan, S. Rein, D. Lucani, and F.H.P. Fitzek. Network Coding in the Bidirectional Cross: A Case Study for the System Throughput and Energy. In *IEEE VTC2013-Spring: Cooperative Communication, Distributed MIMO and Relaying*, Dresden, Germany, June 2013.

[3] M. Hundebøll, J. Leddet-Pedersen, J. Heide, M.V. Pedersen, S.A. Rein, and F.H.P. Fitzek. Catwoman: Implementation and Performance Evaluation of IEEE 802.11 based Multi-hop Networks using Network Coding. In *IEEE VTS Vehicular Technology Conference. Proceedings*. IEEE, 2012.

[4] M. Hundebøll, S.A. Rein, and F.H.P. Fitzek. Impact of Network Coding on Delay and Throughput in Practical Wireless Chain Topologies. In *IEEE CCNC - Wireless Communication Track*, 2013.

[5] W.D. Hamilton. The Evolution of Altruistic Behavior. *The American Naturalist*, 97:354–356, 1963.

[6] F.H.P. Fitzek and M. Katz, editors. *Cooperation in Wireless Networks: Principles and Applications – Real Egoistic Behavior is to Cooperate!* ISBN 1-4020-4710-X. Springer, April 2006.

[7] L. Humphreys. Mobile Social Networks and Social Practice: A Case Study of Dodgeball. *Journal of Computer-Mediated Communication*, 13(1), 2007.

[8] J. P. Onnela, J. Saramäki, J. Hyvönen, G. Szabo, D. Lazer, K. Kaski, J. Kertesz and A.-L. Barabasi. Structure and Tie Strengths in Mobile Communication Networks. *Proceedings of the National Academy of Sciences*, 104(18):7332–7336, 2007.

[9] N. Eagle, A.S. Pentland, and D. Lazer. Inferring Friendship Network Structure by using Mobile Phone Data. *Proceedings of the National Academy of Sciences*, 106(36):15274–15278, 2009.

[10] E. Miluzzo, N.D. Lane, K. Fodor, R. Peterson, H. Lu, M. Musolesi, S.B. Eisenman, X. Zheng, and A.T. Campbell. Sensing meets Mobile Social Networks: the Design, Implementation and Evaluation of the CenceMe Application. In *Proceedings of the 6th ACM Conference on Embedded Network Sensor Systems*, SenSys '08, pages 337–350. ACM, 2008.

[11] A. Beach, M. Gartrell, S. Akkala, J.J. Elston, J. Kelley, K. Nishimoto, B. Ray, S. Razgulin, K. Sundaresan, B. Surendar, M. Terada, and R. Han. WhozThat? Evolving an Ecosystem for Context-aware Mobile Social Networks. *Network, IEEE*, 22(4):50–55, 2008.

[12] L. Hardesty MIT News Office. Secure, Synchronized, Social TV. http://web.mit.edu/newsoffice/2011/social-tv-network-coding-0401.html.

[13] wpmudev. Pay with a Like. https://premium.wpmudev.org/project/pay-with-a-like/.

[14] springwise access. Web page of springwise access. http://www.springwise.com/telecom_mobile/portable-4g-hotspot-rewards-us ers-sharing-mobile-internet-connections/.

[15] F.H.P. Fitzek, J. Heide, M.V. Pedersen, and M. Katz. Implementation of Network Coding for Social Mobile Clouds. *IEEE Signal Processing Magazine*, January 2013.

[16] M.V. Pedersen and F.H.P. Fitzek. Mobile Clouds: The New Content Distribution Platform. *IEEE Transaction on Entertainment Technologies*, May 2012.

[17] C. Wu, M. Gerla, and M. van der Schaar. Social Norm Incentives for Secure Network Coding in MANETs. In *NetCod 2012*, Boston, USA., 2012.

Part Four

Green Aspects of Mobile Clouds

9

Green Mobile Clouds: Making Mobile Devices More Energy Efficient

Smartphones are fun for about half an hour... Then the battery dies.

Alex Smith

Green communications has gained a lot of interest lately and the question here is how mobile clouds can help to improve the energy efficiency of mobile devices. In this chapter the energy-saving potential of mobile clouds is demonstrated. A particular scenario is chosen for this purpose, namely a cooperative download case where users retrieve a desired content in a collaborative manner. The effect on download speed is also investigated. Several cooperative strategies for the mobile cloud are studied and their energy–saving performance is compared.

9.1 Introduction

In the last few years green communications has raised considerable interest in the research community. The carbon footprint of the ICT sector has been discussed intensively and a new research area was born called green communications. The main focus of green communications is not really on the reduction of CO_2 emissions, but it is more motivated by the huge electricity bills cloud providers and network operators are facing these days. Without going into detail explaining the relationship between CO_2 emissions and green communication, we would just like to state that any reduction of energy consumption is always a good idea. The larger the

Mobile Clouds: Exploiting Distributed Resources in Wireless, Mobile and Social Networks, First Edition.
Frank H.P. Fitzek and Marcos D. Katz.

energy saving, the deeper the economic, technical, usability and environmental impacts. From network operators' and service providers' points of view huge energy consumption means prohibitively high energy costs required to maintain their networks and services, and ultimately reductions in their economic profits. On the other hand, from the consumer standpoint, high energy consumption of their mobile devices leads to reduced operational times and eventually to degraded quality of experience. The mobile phones have to support several services, several air interfaces and large displays. In [1, 2] we referred to this as *energy trap* in alignment with the *evolutionary trap* in Chapter 7.

Energy–efficient solutions are also important for telecom (infrastructure) and mobile device manufacturers, as offering low–energy equipment provides them with a very competitive edge for their business. This chapter focuses mainly on the potential of mobile clouds for improving energy efficiency of mobile devices. The next chapter explores the use of mobile clouds for enhancing energy efficiency at the infrastructure side.

The main focus in this chapter is the energy consumption resulting from the wireless communication and the computational efforts associated with operation of the mobile cloud. Operational costs not directly related with communication activities and functionalities are not included in the developed models. In [3, 4] we presented and discussed energy consumption for individual parts and functionalities of mobile phones such as display, CPU, storage access, Bluetooth, Wi–Fi or cellular communications blocks.

Results showed that the wireless communications–related functionalities consume more energy than do other parts on the mobile such as the CPU, the display, or memory access. Based on those findings we will investigate the energy consumption for standalone (e.g., non–cooperative) and mobile cloud–based communication. In order to make the reader familiar with the energy-saving potentials of mobile clouds we first compare a single standalone user case with a mobile cloud case for the download scenarios shown in Figure 9.1 and Figure 9.2, respectively. Here we focus on the download scenario but later will extend it to streaming services as well. The download case represents for instance firmware updates or file download

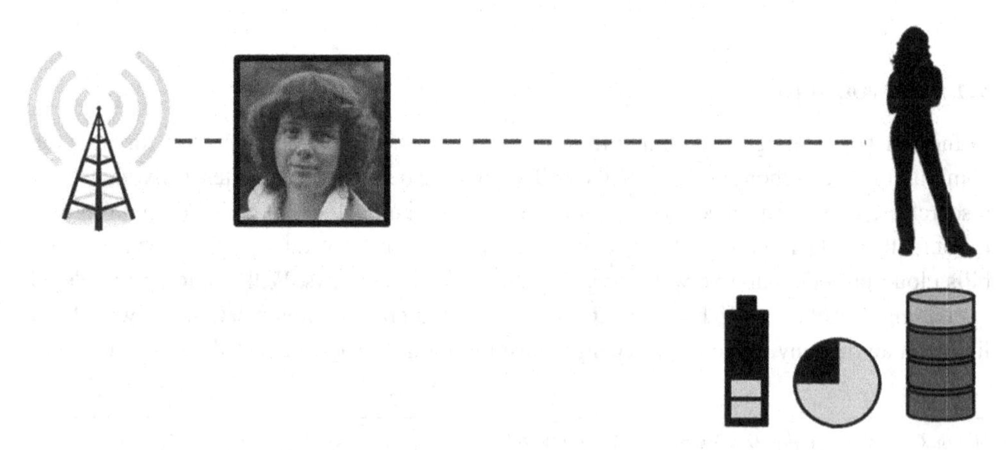

Figure 9.1 Single-user (non–cooperative) case – the user has to retrieve the full information content from the overlay network.

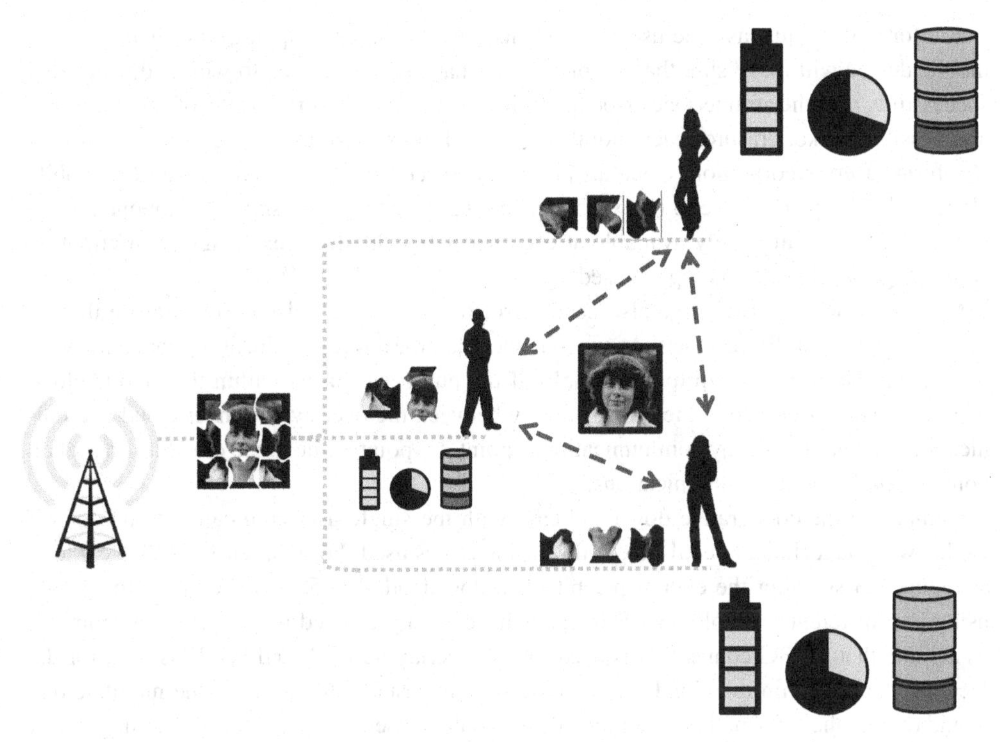

Figure 9.2 Mobile cloud case – cooperative users share tasks to download from the overlay network.

a la BitTorrent. The streaming cases on the other hand target real-time services such as distribution of sport or any other live events. In both cases we assume the download of certain content from the network to the mobile device.

Starting with the single–user case in Figure 9.1, it is quite obvious that the full download, which is represented here by a single photo, has to take place over a single air interface. This download will require a certain time, certain bandwidth share from the overlay network and consumes a certain level of energy on the device side. This reflects the state of the art of today's communication architecture.

In Figure 9.2 the mobile cloud scenario is shown with three cooperative users and their associated mobile devices. All three users still have a connection to the overlay network and at the same time use short–range wireless links to connect to their cooperating partners. This principle was already discussed in previous chapters, but it is just repeated here to highlight that the cooperating partners should be in close proximity to each other while there is no such restriction on the distance between the serving base station and the nodes of the mobile cloud. As the three users are cooperating each will download a part of the overall information. This is illustrated by a piece of a puzzle. It is of utmost importance that the three users will make sure they get disjointed information (in our example represented by different pieces of the puzzle) and do not download duplicates. In the later part of Chapter 5 we explained that network coding can help significantly to overcome the problem of duplicates or overlapping parts.

But for now we just assume users receive disjointed (non–overlapping) information. Each mobile device will make sure that the pieces they have been assigned to will be downloaded successfully over the air interface associated with the overlay network. In case of channel errors and possible packet erasures each mobile device will make sure those parts are corrected by any form of error correction. Once all pieces are successfully downloaded into the mobile cloud nodes the pieces have to be exchanged among nodes to make sure all component parts of the picture are ultimately available at each node. For this local exchange of information short–range wireless technology is used.

Certainly, the local exchange also needs error-recovery mechanisms to guarantee that all pieces are successfully received. As we will show later, even for the local exchange network coding will be of great help too, especially if the number of users within the mobile cloud increases. The difference in the error recovery between the overlay and inner cloud communications is that the overlay communication is point–to–point while in the cloud it is based on point–to–multi–point communications.

Comparing the cooperative download case with the single-user case can only be done if we know some technical details about the air interfaces used. For that reason we have a look at Figure 9.3 showing the energy per bit for a download of a 500kB file on a Nokia N95 using four different technologies. The results have been presented in [3, 4]. At this point let us assume that the 3G connection is used for the overlay network and Wi–Fi is used for the local exchange within the cloud. In this case we can assume that users in the mobile cloud would receive the information faster than the standalone user. At the beginning the single-user case and the mobile cloud would download from the overlay network at a given rate R. In our example of three cooperative users, the mobile cloud would stop after one third of the time

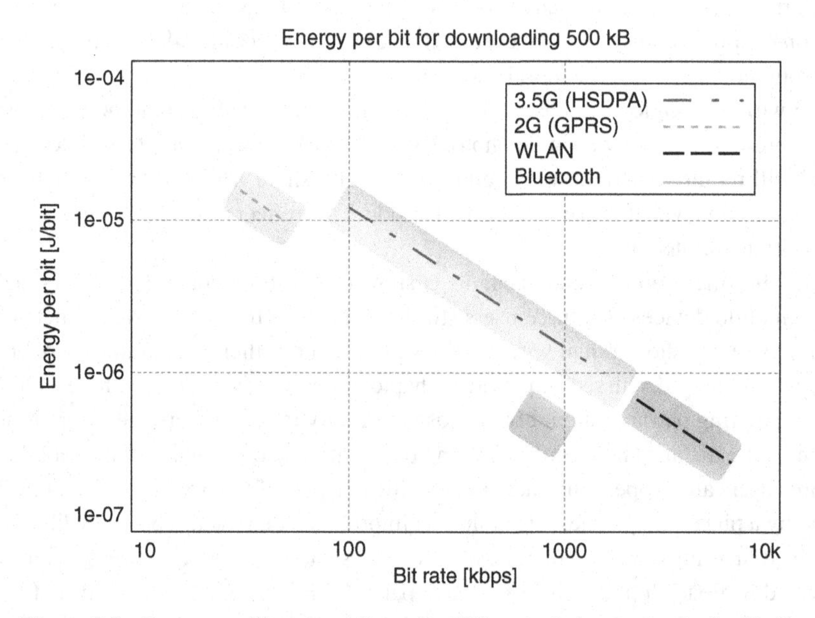

Figure 9.3 Energy per bit versus bit rate for different wireless technologies [3, 4].

the standalone user would take to make the full download. While the single-user case would continue to download with the same rate, the mobile cloud would start to exchange the missing information at a higher rate within the mobile cloud. The mobile cloud would be even faster if the exchange is started in parallel with download from the overlay network.

It is interesting to note that the energy per bit is lower using additional Wi–Fi connections instead of one single 3G connection and we claim that this is the key reason for energy savings in mobile clouds. At this point we do not quantify the energy-saving gains, but we conclude there is a potential for significant energy savings. Concrete evidence will be presented next. The error recovery and the related energy costs for that are also worth considering. Indeed, the more efficient we can handle the error recovery the better will be the performance of the mobile cloud. Not surprisingly network coding will kick in for this case too and we refer to Chapter 5 further discussions on this issue. For the sake of completeness we list the other technology combinations too:

1. 2G / Wi–Fi: In this case the gain provided by the mobile cloud will be even larger as the data rate supported by the 2G connection is smaller than that of 3G, which results in a longer transfer delay for the standalone user as well as an increase of the energy/bit ratio. An increased energy/bit ratio is in favor in the mobile cloud as each node uses the 2G connection only for a fractional time compared with the standalone node. Therefore, the delay reduction and energy savings are improved by the mobile cloud.
2. 2G / Bluetooth: Owing to the smaller data rate on the exchange within the cloud the energy saving gain is smaller compared with the 2G/Wi–Fi case.
3. 3G / Bluetooth: This combination is interesting as the data rate of the overlay network could be in the best of the cases equal to or higher than those on the Bluetooth links. Therefore, it will be hard to achieve delay reduction for the mobile cloud though energy savings are still possible.

On a practical note, Bluetooth has one major drawback for the mobile cloud and this is the absence of a broadcast channel. Even though some Bluetooth chip sets allow the use of broadcast for the Bluetooth master, slaves never have the chance to broadcast. We have shown in [5] that this could lead to significant performance degradation for the mobile cloud in case there are more than three users cooperating in the mobile cloud. Therefore, we will only consider Wi–Fi in this chapter from here on. The interested reader is referred to [5] for a full discussion about Bluetooth technology for mobile clouds.

A second note regards the evolution of data rates as given in Chapter 4. While we see a steady data-rate increase in the cellular systems rolling out 4G technologies worldwide, the commercial roll out of new and faster short–range technologies has nearly stopped for smart phones. Our first implementation of cooperative scenarios was based on GPRS and Bluetooth [6] technologies. As we will see later the ratio of the data rates between the overlay network and the short–range network will determine the potential energy-saving gain of the mobile cloud.

Nevertheless, even though not yet realized it is clear that the data rate for short–range communication will be always be larger than the overlay network, especially if we distinguish

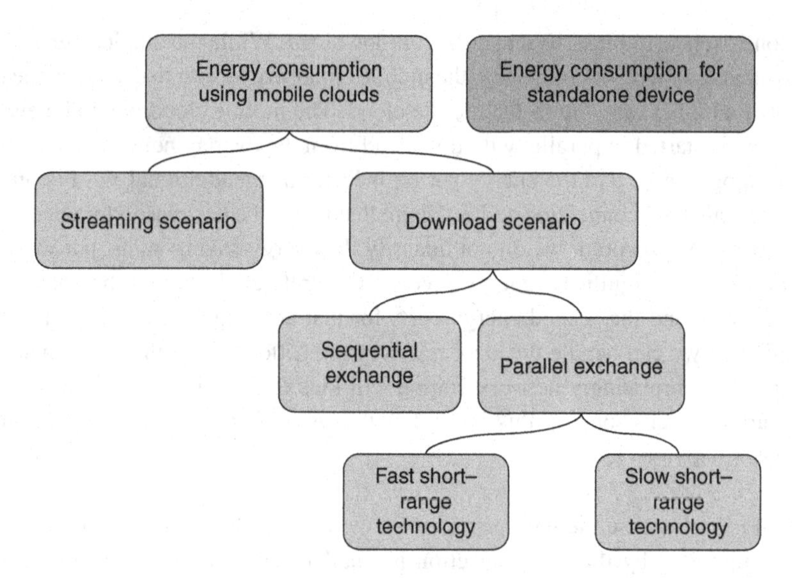

Figure 9.4 Structure of the chapter for different scenarios.

between the data rates that can be physically achieved and those that will be available on the IP level. Currently we see that the overlay networks are more and more loaded and single users are always getting a lesser portion of the overall available data rate. In the following we will go through some examples for two different services for mobile clouds, namely cooperative download and cooperative streaming as given in Figure 9.4. The download scenario has two different rationalization forms, namely the sequential exchange and the parallel exchange. The energy consumption differs in dependency of the short–range technology and we are looking therefore in two more sub cases. After explaining the basics for the individual scenarios we will derive some simple equations for the energy-saving potentials. The energy consumption of a standalone mobile device is used as a baseline to compare the energy improvements achieved by the mobile cloud concept.

9.2 Cooperative Download

The first scenario is called cooperative download. We have previously discussed this scenario to highlight some key mobile cloud ideas but here we will further elaborate on it, deriving the energy consumption of the mobile cloud.The basic setup is given in Figure 9.5. As an example we assume four mobile devices that would like to download the same content from a server. Each device is connected directly to the overlay network using the cellular link and in parallel with all three other devices using short–range links. The overlay network communicates with the devices in a unicast fashion.

As shown in Figure 9.5 the communication takes place in parallel and both air interfaces, the cellular and the short–range ones, are activated at the same time. For the short–range link we assume that the devices can broadcast the information to the cooperating partners in the

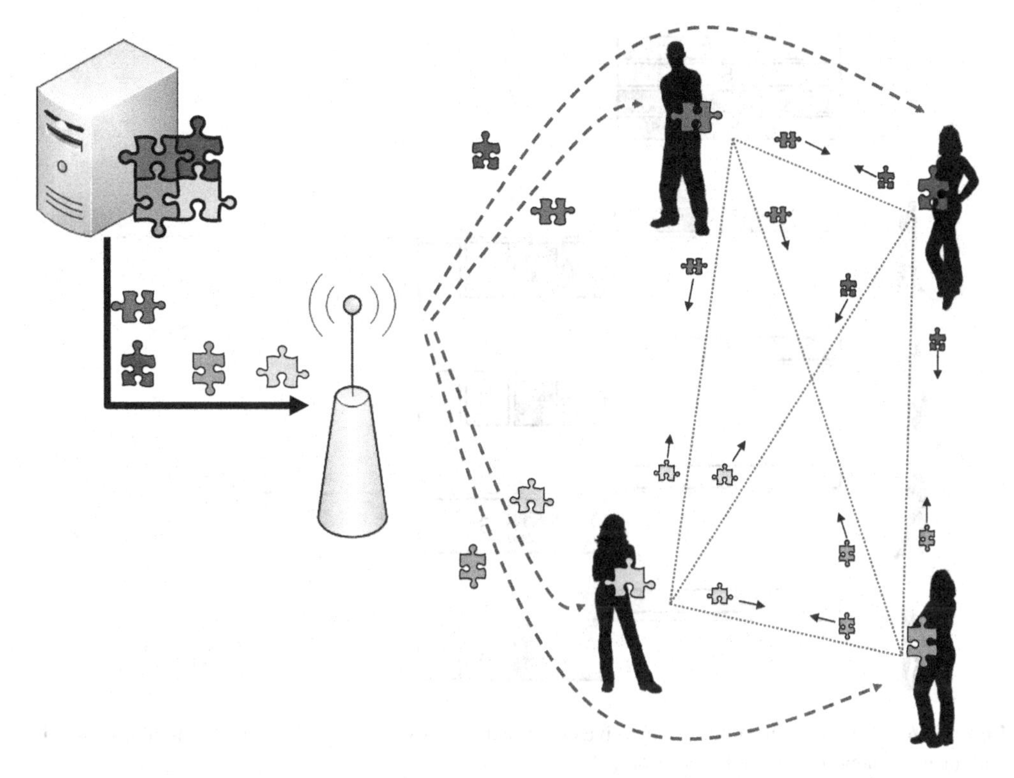

Figure 9.5 Cooperative download scenario with cellular unicast links to the users.

mobile cloud. For the time being we do not consider packet losses, neither on the cellular nor on the short–range links. As we are considering cooperative downloading the server will send in parallel disjointed (non–overlapping) partial information to the four mobile users. The cooperative content distribution is illustrated here by four different pieces of a puzzle. Each mobile user will receive only one unique piece of the puzzle via the cellular link. We underline again that it is of utmost importance that each mobile device receives a different piece in order to avoid undesired data duplication or distribution of incomplete content. Once the pieces have been received by the involved mobile devices, they will locally exchange them. In other words, each mobile device will send out its own piece of the puzzle received from the cellular link. After that procedure each mobile device will have all the four required pieces of the puzzle.

Figure 9.6 shows an activity chart of the cooperative download strategy exploited by the mobile cloud. First a parallel transmission from the network towards the mobile devices takes place (upper part of Figure 9.6) and then each device sends its own content once and receives three pieces over short–range links (lower part of Figure 9.6). As we assume a higher data rate for the short–range communication. receiving of a piece of the puzzle via the short–range will take less time than receiving it via the cellular link. This fact is reflected by the shorter bars on the short–range link in the activity chart even though the same amount of data is transported. Clearly, the time a standalone mobile device would take get the complete information is four

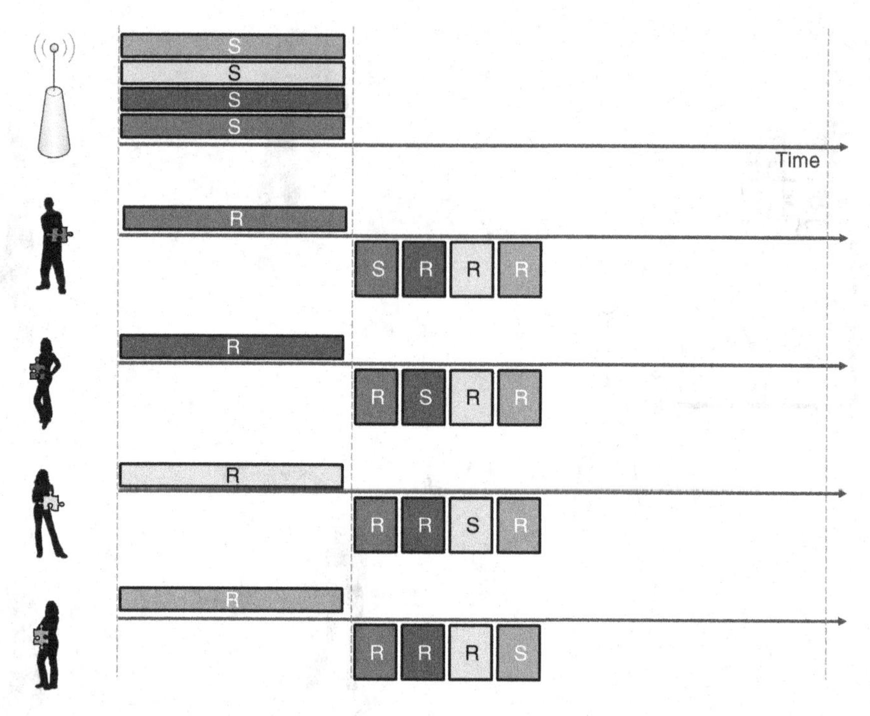

Figure 9.6 Activity chart of the cooperative download scenario (sequential order of cellular download and local exchange) discussed in Section 9.2.1.

times as long as the activity on the cellular link shown in Figure 9.6 as delivering the four pieces takes place sequentially.

Figure 9.6 shows that the time needed for the cooperative download by the mobile cloud is given by the time for the partial cellular download (one fourth of the standalone time) in addition to the time needed for the local exchange, which only can be started after the content is available at the devices.

If the amount of data that is downloaded is large, the cellular download can be done in batches, so that after each batch the local exchange will take place in parallel to the download of the next batch, as illustrated by Figure 9.7. If the number of batches is large enough the download time for the mobile cloud with four cooperative mobile devices is just one fourth of the time a standalone user would have taken. We will come back to this example later in this chapter. After describing the cooperative download scenario we derive the energy consumption for the standalone and the cooperative case. In general the energy E equals the product of power level P and the time t. For the standalone case the mobile device is just receiving the data from the overlay network such that the energy spent E_{over} for the download equals the download time t_c and the power level (i.e., power consumption) for receiving from the overlay network P_c as given in Equation 9.1.

$$E_{over} = P_c \cdot t_c \tag{9.1}$$

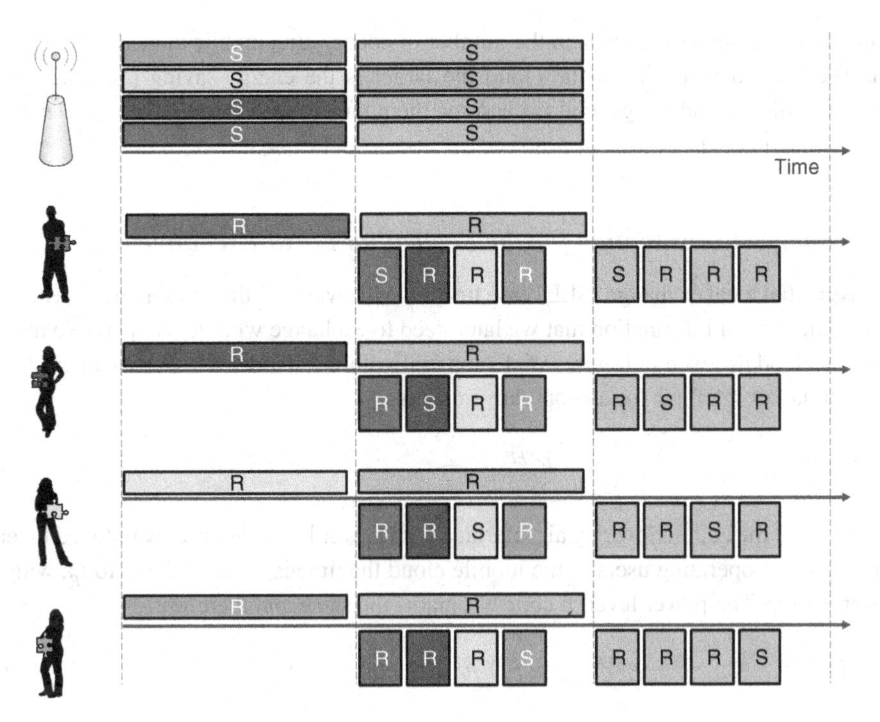

Figure 9.7 Activity chart of the cooperative download scenario (parallel cellular download and local exchange) discussed in Section 9.2.2.

The download time t_c is given by the amount of data D that a user wants to download divided by the available data rate of the overlay system R_c. The power level P_c can be obtained from the data sheet of the chosen wireless technology (see Chapter 4). For the cooperative case the calculation of the energy E_{coop} becomes more cumbersome. Before we proceed with that calculation we define the energy gain G as given

$$G = E_{coop}/E_{over}. \tag{9.2}$$

G reflects the percentage of energy that will be used compared with the standalone case. Now let us look into the energy E_{coop} consumed by each user in the mobile cloud. First of all we have to distinguish two main cases, namely sequential or parallel local exchange. These two main cases were already introduced by Figure 9.6 and Figure 9.7. In the sequential case all mobile devices download the information assigned to them and then start the local exchange. In the parallel case the mobile devices get their information in batches (e.g., a single IP packet) and exchange it in parallel. The main difference is that for the sequential case only one air interface at the time is active, while for the parallel case two air interfaces are simultaneously active. These two cases have a significant impact on the energy consumption and the delay as we will see later.

For both cases, the overall energy E_{coop} is the sum of the energy spent on the cellular air interface and the short–range interface. On the cellular air interface the energy spent for one individual mobile device depends on the number of cooperating mobile devices in the mobile cloud. The more devices are in the cloud the larger is the energy-saving potential for the mobiles. In order to make it general we assume the number of cooperative mobile devices in a mobile cloud to be J.

9.2.1 Energy Consumption for the Sequential Local Exchange (SLE)

In the sequential local exchange (SLE) case first we will switch on the cellular air interface and download the partial information that we later need to exchange with the cooperative users in the mobile cloud as given in Figure 9.6. Compared with the standalone user the energy spent on the cellular air interface for a cooperative user is

$$E_{coop,C,r}^{SLE} = t_p \cdot P_{C,r}. \tag{9.3}$$

The usage of the cellular overlay air interface with power level $P_{C,r}$ is now reduced in terms of time. For J cooperating users in the mobile cloud the time is reduced down to t_p, which is one over J of t_c. The power level of course remains the same and therefore

$$E_{coop,C,r}^{SLE} = t_p \cdot P_{C,r} = 1/J \cdot t_c \cdot P_{C,r}. \tag{9.4}$$

When the partial cellular download is completed the overlay air interface can be shut down and the local short–range air interface will be switched on. The local exchange will start with a sending phase followed by a receiving phase (this is for illustrative purposes only, as some mobile device in the cloud will send its own content before and/or after receiving data from the cloud members). The energy for sending on the short–range for the mobile cloud members is

$$E_{coop,SR,s}^{SLE} = t_{sr,s} \cdot P_{sr,s} = \frac{1}{JZ} \cdot t_c \cdot P_{sr,s}, \tag{9.5}$$

with Z being the ratio between the short–range data rate and the cellular data rate. $P_{sr,s}$ is the power level of the short–range technology in the sending mode. The time the user has to send is denoted by $t_{sr,s}$. Compared with the time t_c for receiving the full data, $t_{sr,s}$ is smaller by the number of the cooperating devices but is also reduced due to the larger data rate available on the short–range links compared with the cellular links. Actually the fraction of time for sending on the short–range compared with the cellular download is $1/(JZ)$. Now we have to calculate the energy for receiving part in the mobile cloud which equals

$$E_{coop,SR,r}^{SLE} = t_{sr,r} \cdot P_{sr,r} = \frac{J-1}{JZ} \cdot t_c \cdot P_{sr,r}. \tag{9.6}$$

$P_{sr,r}$ is the power level of the short–range technology in the receiving mode. And assuming that the short–range receiving and sending data rate is the same, the time for receiving is $J-1$ times larger than the sending. This is obvious as the cooperating device is contributing $1/J$

of the full data D and needs to receive $1 - 1/J$ from the cooperating devices. Before we sum up the different parts to calculate the overall energy E_{coop}, we would like to note that the energy is just dependent on technology parameters such as the power levels and the data rates resulting in the parameter Z as well as the scenario-dependent value J. Now the overall energy consumed for a single user in the mobile cloud for the sequential local exchange (SLE) is

$$E_{coop}^{SLE} = E_{coop,C,r}^{SLE} + E_{coop,SR,s}^{SLE} + E_{coop,SR,r}^{SLE}. \tag{9.7}$$

Using previous results 9.3, 9.5, and 9.6 it becomes

$$E_{coop}^{SLE} = \frac{1}{J} t_c P_c + \frac{1}{JZ} t_c P_{sr,s} + \frac{J-1}{JZ} t_c P_{sr,r}. \tag{9.8}$$

The overall time for the full exchange to be completed is

$$T_{coop}^{SLE} = 1/J \cdot t_c + 1/Z \cdot t_c. \tag{9.9}$$

It is just calculated by summing up the time for the time on the cellular air interface $(1/J \cdot t_c)$ and the time for the local exchange $(J \cdot 1/JZ \cdot t_c)$.

Figure 9.8 depicts the energy gain G as a function of the number of collaborative users in a mobile cloud. For the results achieved we used the parameters shown in Table 9.1. Those parameters will be used throughout this chapter in order to be able to compare the performance figures of the different scenarios with each other. For each value of J different energy parts of the overall energy are shown. In the bottom the energy for receiving on the overlay network,

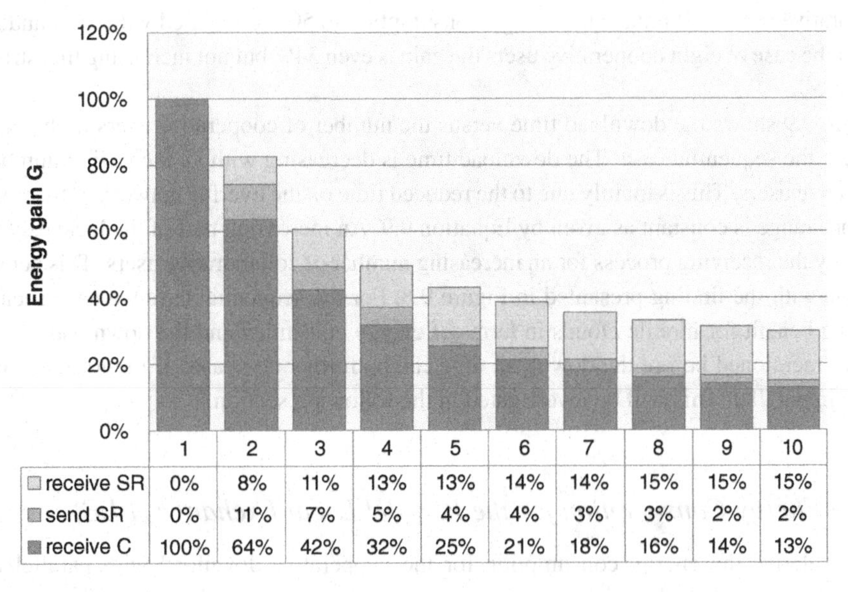

	1	2	3	4	5	6	7	8	9	10
receive SR	0%	8%	11%	13%	13%	14%	14%	15%	15%	15%
send SR	0%	11%	7%	5%	4%	4%	3%	3%	2%	2%
receive C	100%	64%	42%	32%	25%	21%	18%	16%	14%	13%

Figure 9.8 Energy gain G for the sequential case (SLE) versus the number of cooperative users in a mobile cloud split up into the energy consumed on cellular and short–range.

Table 9.1 System parameters used throughout this chapter.

System parameters name	Power consumption
$P_{C,r}$	1.1 W
$P_{C,i}$	0.1 W
$P_{sr,s}$	1.4 W
$P_{sr,r}$	1.1 W
$P_{sr,i}$	0.1 W
Z	6

in the middle the energy to send, and on the top the energy to receive within the mobile cloud are depicted. From Figure 9.8 we see that the overall energy consumption is reduced with an increasing number of cooperative users (i.e., cloud size). We also see that the energy for receiving on their overlay network as well as the energy to send within the mobile cloud decreases as the cloud size increases. The reason behind this is obvious as more cooperative users are able to split the common content into more pieces. As each device is receiving less data from the overlay with an increasing number of cooperative users, also the sending activity is proportionally reduced. On the other side the missing parts of the common content need to be gathered within the mobile cloud and the energy consumed for receiving with the mobile is increasing for a larger number of users. For example, for ten cooperative users the energy for receiving in the mobile cloud (15%) exceeds the overall energy spent on receiving for the overlay network and the energy to send in the mobile cloud.

Figure 9.8 shows also the fact that the largest improvement of the gain for the mobile cloud in terms of energy savings is achieved with the very first collaborative users. For instance, four collaborative users will reduce the energy consumption to 50% compared with the standalone case. In the case of eight cooperative users the gain is even 34% but not increasing that strongly anymore.

Figure 9.9 shows the download time versus the number of cooperative users in the mobile cloud for the sequential case. The download time is decreasing with an increasing number of cooperative users. This is mainly due to the reduced time on the overlay network as the time on the short–range is constant as given by Equation 9.9. An increasing part of the local exchange is used by the receiving process for an increasing number of collaborative users. This behavior is in line with the finding presented in Figure 9.8. For the sequential case we have seen that there is a benefit for mobile clouds in terms of energy consumed and the download time. As we have mentioned before the download time can be further decreased if the local exchange is done in parallel. This will be investigated in the following section.

9.2.2 *Energy Consumption for the Parallel Local Exchange (PLE)*

Here we derive the energy consumption for the cooperative download with parallel local exchange. As mentioned earlier, we will consider two cases exploiting fast and slow short–range technologies. To be more precise the fast case assumes that the exchange on the

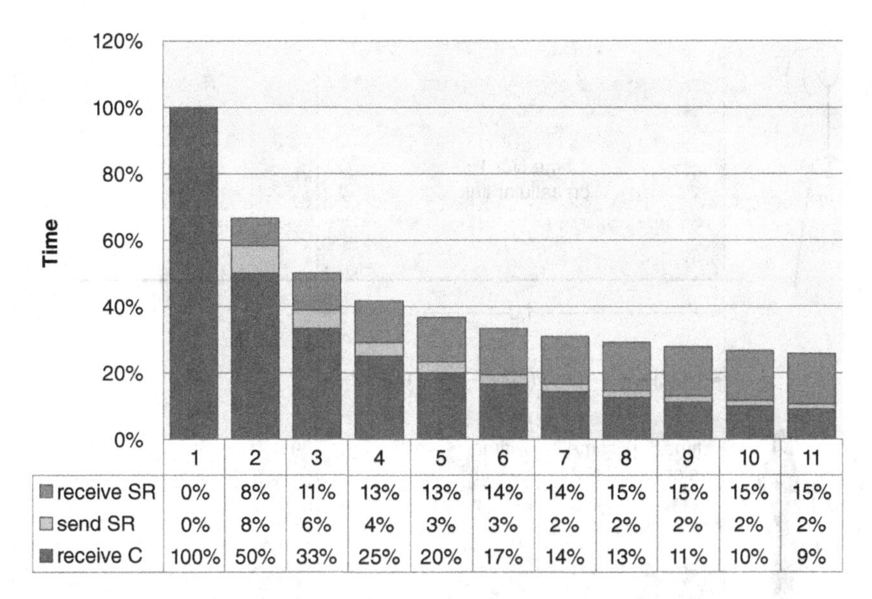

Figure 9.9 Download time for the sequential case (SLE) versus the number of cooperative users in the mobile cloud split up into the time used on cellular and short–range.

short–range can be done faster than the retrieval of the partial information on the overlay network. With an increasing number of cooperative devices J in a mobile cloud, the time for the partial information retrieval on the overlay network shrinks, while the exchange phase remains constant. We start in Section 9.2.2 with the fast short–range technology and later consider the case of the slow short–range technology.

Fast Short–Range Technology

As given in Figure 9.10 the download time on the cellular air interface is reduced to

$$T_{coop}^{PLE,fast} = 1/J \cdot t_c, \tag{9.10}$$

assuming that the full data retrieval would need the t_c. Therefore the energy spent on the cellular air interface equals

$$E_{coop,C,r}^{PLE,fast} = 1/J \cdot t_c \cdot P_{C,r}. \tag{9.11}$$

On the short–range link the energy used is composed of three activity phases as given in Figure 9.10. The first activity part is the sending phase followed by the receiving phase and the idle phase. The power levels for sending, receiving and idle are technology dependent but in general it can be assumed that $P_{sr,s} > P_{sr,r} > P_{sr,i}$ (see Table 9.1). In order to calculate the time a mobile will spend in one of the three phases we reuse the parameter Z which equals the ratio of the data rate on the short–range to the data rate of the cellular air interface.

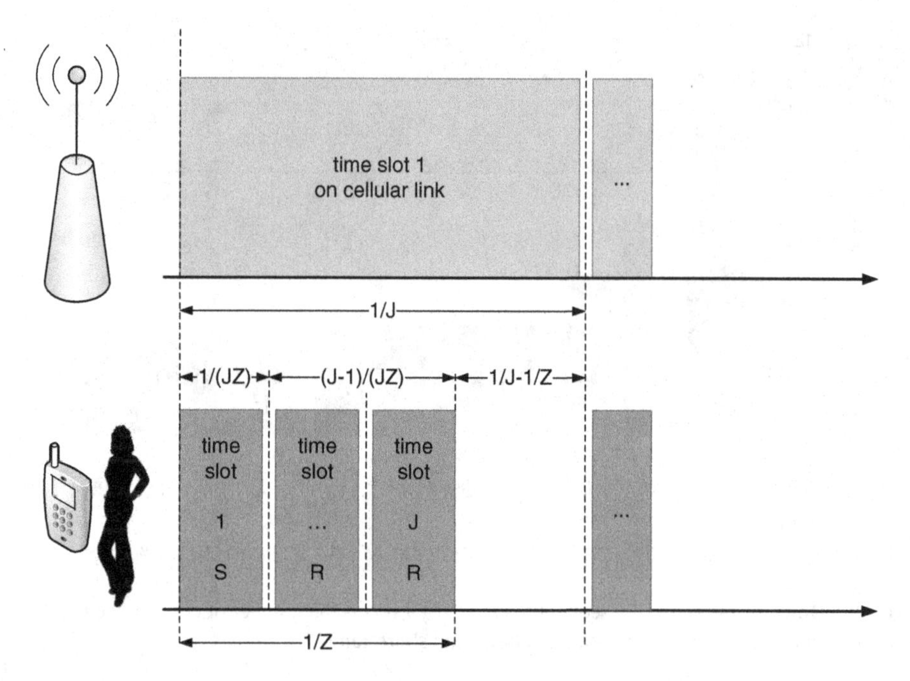

Figure 9.10 Cooperative download for the cellular and the short–range scenario (1/*J* > 1/*Z*).

Now we are able to give the time the mobile device is in a certain phase on the short–range. The time $1/J$ was used for receiving the information on the cellular link and therefore the time $1/JZ$ will be used to relay it on the short–range link.

$$E_{coop,SR,s}^{PLE,fast} = t_{sr,s} \cdot P_{sr,s} = \frac{1}{J \cdot Z} \cdot t_c \cdot P_{sr,s}. \tag{9.12}$$

As we received only one piece of the puzzle, the other pieces, to be more accurate the missing $J - 1$ pieces, need to be received over the short–range link. As we assume to have the same data rate for sending and receiving, the time in the receiving state equals $(J - 1)/JZ \cdot t_c$. Now the exchange of the data of the last batch is completed, but reception on the cellular link for the new batch is still taking place so that the short–range is idle for some time. The time for sending and receiving on the short–range is the sum of $1/J \cdot t_c$ and $(J - 1)/JZ \cdot t_c$ which equals $1/Z \cdot t_c$. This is not surprising as $1/J \cdot t_c$ represents the time to get one piece and t_c represents the time to get all J pieces.

$$E_{coop,SR,r}^{PLE,fast} = t_{sr,r} \cdot P_{sr,r} = \frac{J-1}{JZ} \cdot t_c \cdot P_{sr,r}. \tag{9.13}$$

But as the short–range is Z times faster the activity time of the short–range is $1/Z \cdot t_c$. Now the idle time on the short–range is just $(1/J - 1/Z) \cdot t_c$.

$$E_{coop,SR,i}^{PLE,fast} = t_{sr,i} \cdot P_{sr,i} = \left(\frac{1}{J} - \frac{1}{Z}\right) \cdot t_c \cdot P_{sr,i}. \tag{9.14}$$

Following this small argumentation the overall energy can be calculated as given below by summing up the product of the times in a certain state with its related power values.

$$E_{coop}^{PLE,fast} = E_{coop,C,r}^{PLE,fast} + E_{coop,SR,s}^{PLE,fast} + E_{coop,SR,r}^{PLE,fast}. \tag{9.15}$$

Using Equations 9.11 (assuming now idle time at the cellular air interface), 9.12, 9.13, and 9.14 the overall energy consumption for a cooperative mobile cloud becomes

$$E_{coop}^{PLE,fast} = \underbrace{\frac{1}{J}}_{t_{c,r}} t_c P_{c,r} + \underbrace{0}_{t_{c,i}} t_c P_{c,i} + \underbrace{\frac{1}{JZ}}_{t_{sr,s}} t_c P_{sr,s} + \underbrace{\frac{J-1}{JZ}}_{t_{sr,r}} t_c P_{sr,r} + \underbrace{\left(\frac{1}{J} - \frac{1}{Z}\right)}_{t_{sr,i}} t_c P_{sr,i}. \tag{9.16}$$

In the following section we look into the case when the short–range technology would need more time for the exchange.

Slow Short–Range Technology

In the aforementioned formulation we assumed that the short–range air interface was faster with its exchange than the cellular in downloading one piece of the puzzle. This is given for the case $1/J > 1/Z$. For the sake of completeness we give here the equation if $1/J < 1/Z$ as given in Figure 9.11. As we can see in Figure 9.11 there is no idle time in the short–range phase anymore ($E_{coop,SR,i}^{PLE,slow} = 0$) as the cellular air interface has to wait for the short–range air interface to complete its tasks. But on the other hand, there is now an idle phase on the cellular air interface that needs to be taken into account.

Equation 9.18 calculates the time required for the short–range exchange. As the partial download takes less time, the idle time can be calculated as given in Equation 9.17. The time of the idle phase on the cellular air interface equals $(1/Z - 1/J) \cdot t_c$, which is simply the result of the activity on the short–range air interface subtracted by the receiving activity on the cellular air interface.

$$E_{coop,C,i}^{PLE,slow} = \left(\frac{1}{Z} - \frac{1}{J}\right) \cdot t_c \cdot P_{C,i} \tag{9.17}$$

$$T_{coop}^{PLE,slow} = 1/Z \cdot t_c \tag{9.18}$$

$$E_{coop}^{PLE} = \underbrace{\frac{1}{J}}_{t_{c,r}} t_c P_{c,r} + \underbrace{\left(\frac{1}{Z} - \frac{1}{J}\right)}_{t_{c,i}} t_c P_{c,i} + \underbrace{\frac{1}{JZ}}_{t_{sr,tx}} t_c P_{sr,tx} + \underbrace{\frac{J-1}{JZ}}_{t_{sr,r}} t_c P_{sr,r} + \underbrace{0}_{t_{sr,i}} t_c P_{sr,i} \tag{9.19}$$

Figure 9.12 shows the normalized energy consumption versus the number of cooperating mobile devices with the contribution of the cellular and the short–range technology. With five cooperating devices in the cloud the idle time on the short–range technology disappears. Only with eight cooperating devices does the energy for the local exchange become larger than the

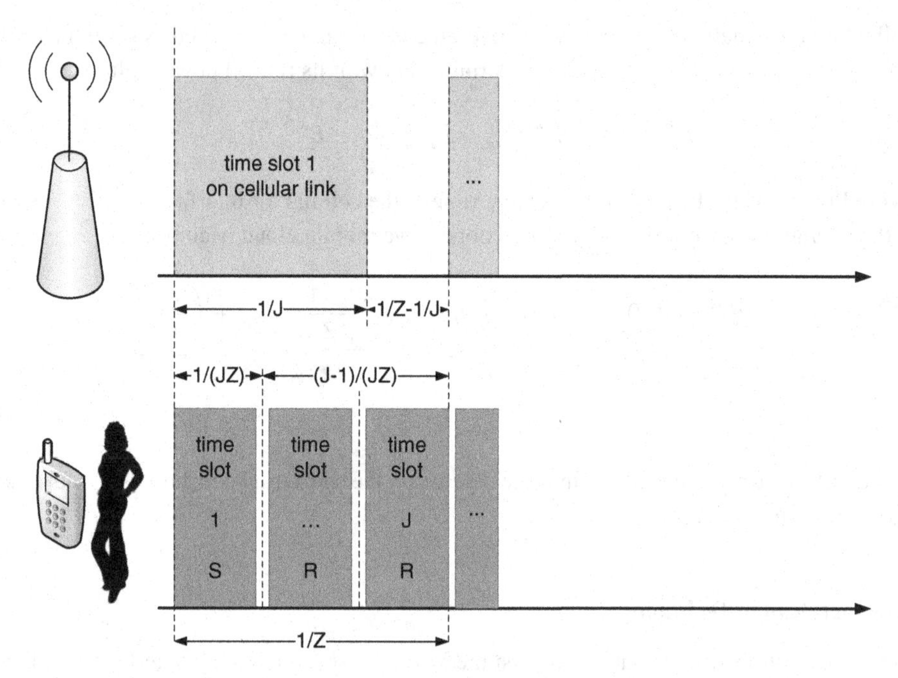

Figure 9.11 Cooperative download for the cellular and the short–range scenario ($1/J < 1/Z$).

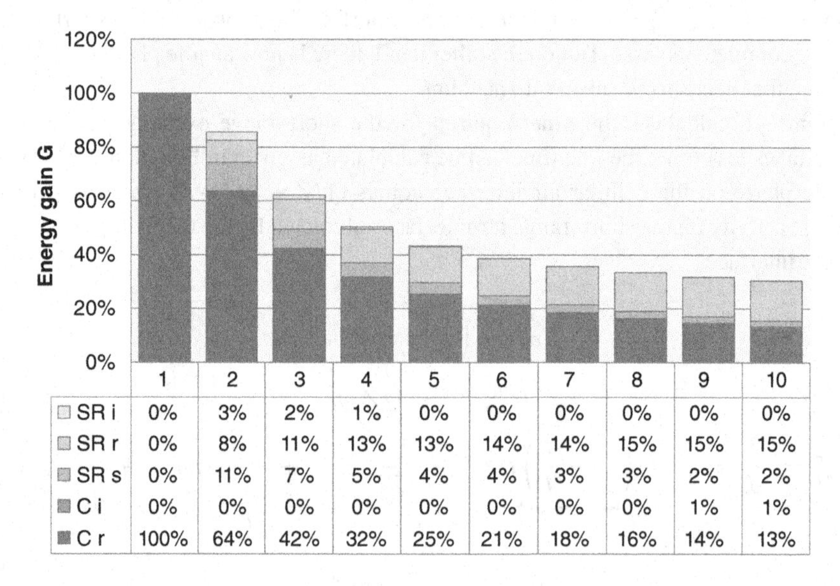

	1	2	3	4	5	6	7	8	9	10
SR i	0%	3%	2%	1%	0%	0%	0%	0%	0%	0%
SR r	0%	8%	11%	13%	13%	14%	14%	15%	15%	15%
SR s	0%	11%	7%	5%	4%	4%	3%	3%	2%	2%
C i	0%	0%	0%	0%	0%	0%	0%	0%	1%	1%
C r	100%	64%	42%	32%	25%	21%	18%	16%	14%	13%

Figure 9.12 Normalized energy consumption versus number of cooperating mobile devices for the parallel local exchange (PLE).

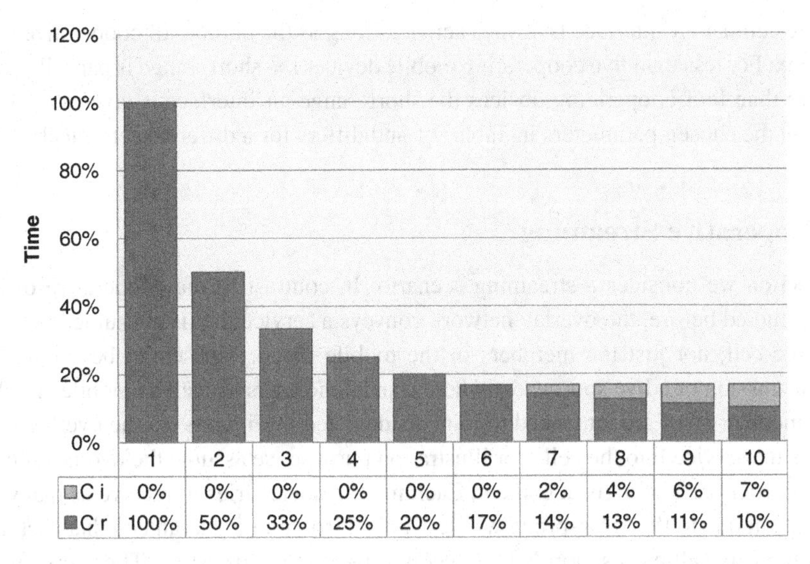

	1	2	3	4	5	6	7	8	9	10
C i	0%	0%	0%	0%	0%	0%	2%	4%	6%	7%
C r	100%	50%	33%	25%	20%	17%	14%	13%	11%	10%

Figure 9.13 Normalized download time versus number of cooperating mobile devices from the cellular air interface's point of view.

cellular part. With a few cooperating devices the dominant part is still the cellular air interface. But this cannot be considered as a general statement, but an interesting outcome resulting from the parameters chosen in Table 9.1.

In Figures 9.13 and 9.14 the normalized download time for the cellular and the short–range air interface are given. Note that both figures show the same times but with different activity

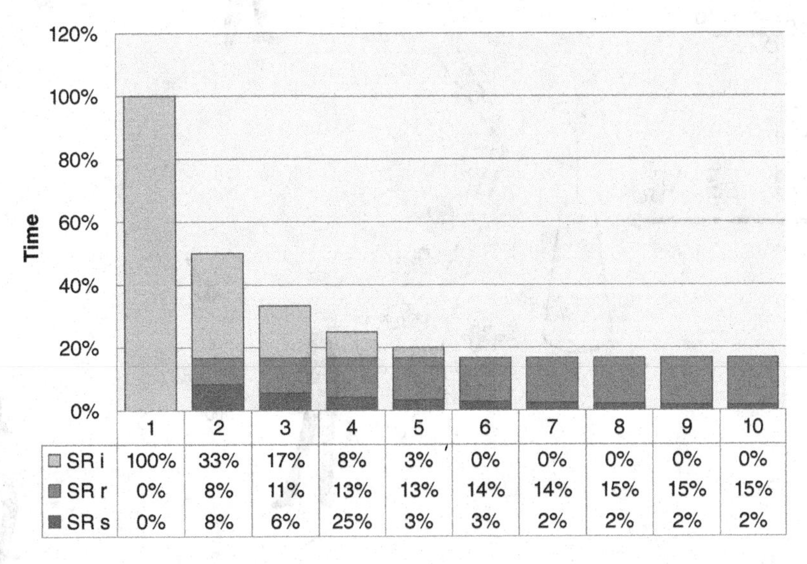

	1	2	3	4	5	6	7	8	9	10
SR i	100%	33%	17%	8%	3%	0%	0%	0%	0%	0%
SR r	0%	8%	11%	13%	13%	14%	14%	15%	15%	15%
SR s	0%	8%	6%	25%	3%	3%	2%	2%	2%	2%

Figure 9.14 Normalized download time versus number of cooperating mobile devices from the short–range air interface's point of view.

levels. The cellular air interface is always active as long as the number of cooperative nodes is less than six. For less than five cooperating mobile devices the short–range is partially inactive. With more than four cooperating devices the short–range air interface is always active. This depends on the chosen parameters in Table 9.1 and differs for a different set of technologies.

9.3 Cooperative Streaming

In this section we consider a streaming scenario. In contrast to the cooperative download approach studied before, the overlay network conveys a service that is consumed by multiple users in the cell, not just the members of the mobile cloud. This could be for instance a video transmission of a live sport event where standalone users as well as mobile clouds with different member sizes are interested to join. As depicted in Figure 9.15 the overlay network broadcasts the service into the cell. For illustration purpose we assume the overlay network is sending out four parts of a given content. Each mobile device tries to receive as many pieces as possible but due to the error–prone wireless links it cannot be guaranteed that each mobile device will receive all pieces, nor that the received pieces are error–free. Therefore, members of the mobile cloud will help each other to recover from the losses. As we will see later the local error recovery will help the overlay network to reduce the amount of redundancy that

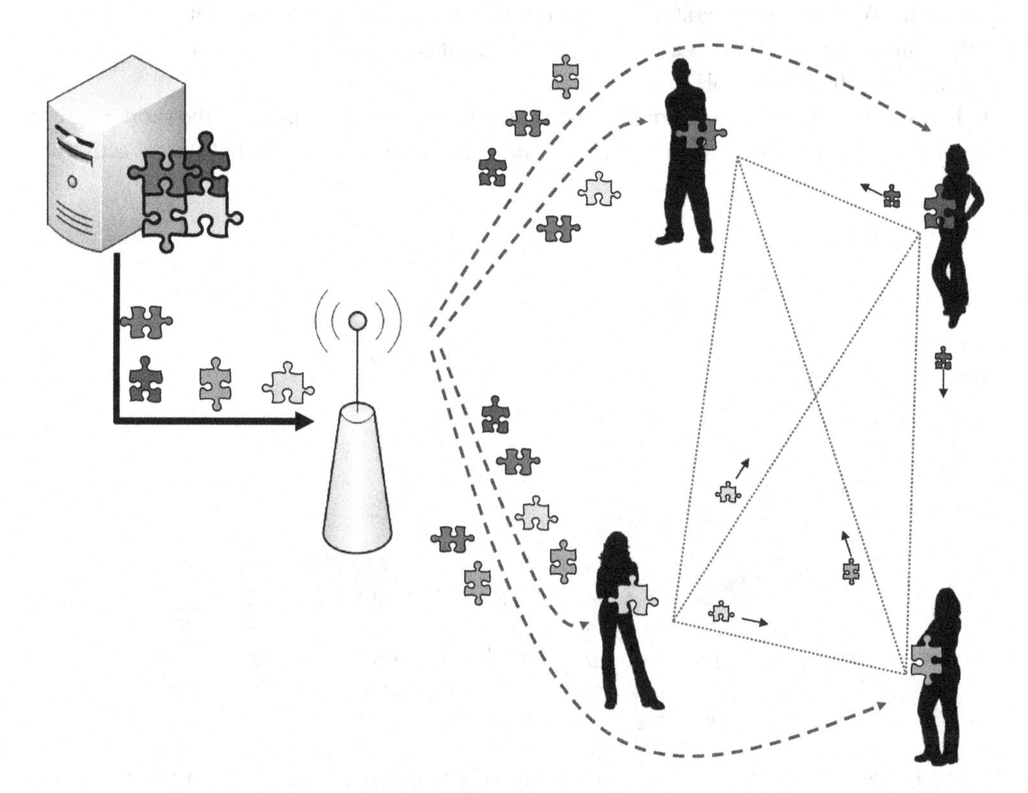

Figure 9.15 Streaming scenario with local error recovery.

it would need to add in case of standalone users. We have considered this in more detail in Chapter 5. In Figure 9.15 the four users experience different losses but as given in the example the mobile cloud as a group has fully received the information and is just relying in local exchange of information. In our example in Figure 9.15 three packets are broadcast within the mobile cloud in order to complete the full information for all four members.

In order to increase the energy-saving potential we can reduce the number of receiving devices significantly. For example, in Figure 9.15 we see that several pieces of the content are received by multiple devices over the air interfaces. And we learned from the examples before that it would be beneficial to reduce the number of receiving activities and switch to idle mode or even to switch off the whole RF/BB chain. Figure 9.16 shows the optimal reception of the streaming content if no errors occur. In this case each device would just tune in for one piece of the content and distribute it later on the short–range air interface. While one device receives the information for the cloud, the other members will be turning off the overlay interface or switching it to idle mode to save energy. The missing pieces, as we have seen before in the download scenarios, can then be received over the short–range interface as well. In case of errors the number of devices listening to a certain piece of the content can be increased. It

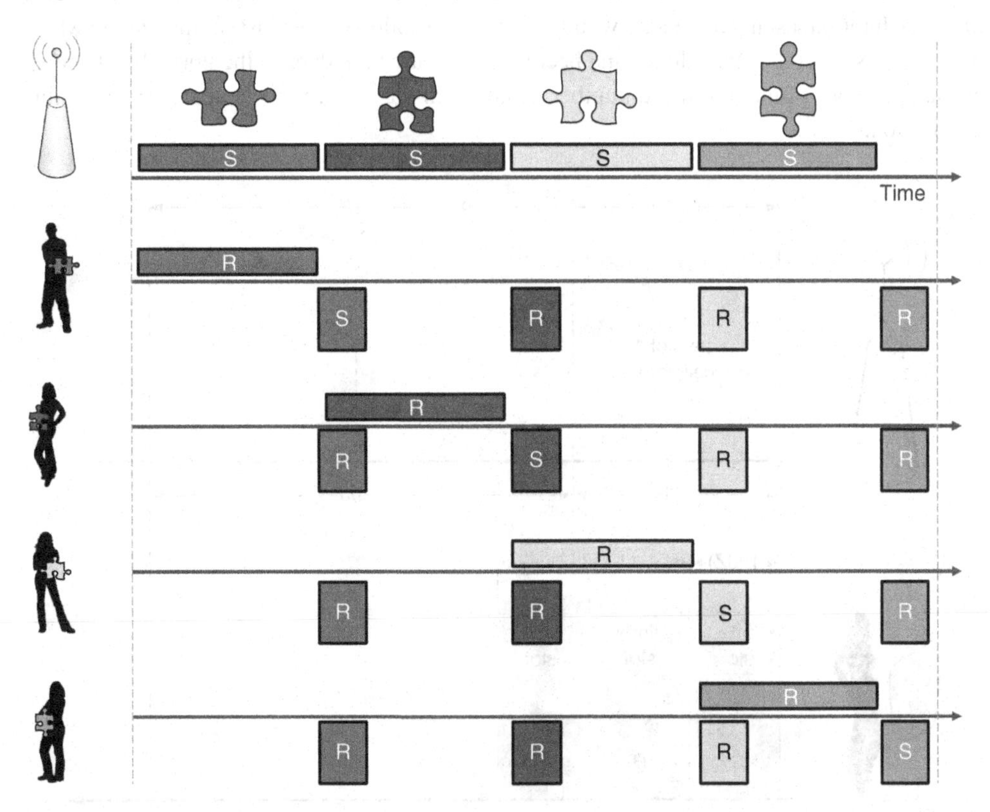

Figure 9.16 Optimal receiving pattern in terms of energy saving in the mobile cloud for the streaming scenario.

will be a trade–off between energy saving and delay. If too many devices are trying to receive a certain piece then it is ensured that the piece is within the cluster but this would lead to increased energy consumption. If the number of devices trying to get the piece is too small and none of the devices is getting it, then the overlay network has to repeat the transmission, which in turn will increase the delay.

$$E_{coop}^{stream} = \underbrace{\frac{1}{J}}_{t_{c,r}} t_c P_{c,r} + \underbrace{\left(1 - \frac{1}{J}\right)}_{t_{c,i}} t_c P_{c,i} + \underbrace{\frac{1}{JZ}}_{t_{sr,tx}} t_c P_{sr,tx} + \underbrace{\frac{J-1}{JZ}}_{t_{sr,r}} t_c P_{sr,r} + \underbrace{\left(1 - \frac{1}{Z}\right)}_{t_{sr,i}} t_c P_{sr,i} \tag{9.20}$$

As given in Figure 9.16 we cannot expect any speed-up in the data reception, which would be even strange as those services typically broadcast live events. So in contrast to the download scenarios where we had a speed-up for the mobile cloud, here we will not see any. To derive the energy consumption for the streaming scenario, we have a closer look at the activity chart given in Figure 9.17. On the air interface for the overlay network we have two phases. The first phase is a receiving phase followed by the second phase being idle. The time duration of one mobile device of a mobile cloud with J cooperative partners is again t/J, with t being the time duration a standalone user would need. In the following we just normalize the whole timing plot so that the standalone time becomes 1. Figure 9.18 depicts the normalized energy consumption as a function of the number of nodes in the cloud for a cooperative streaming application.

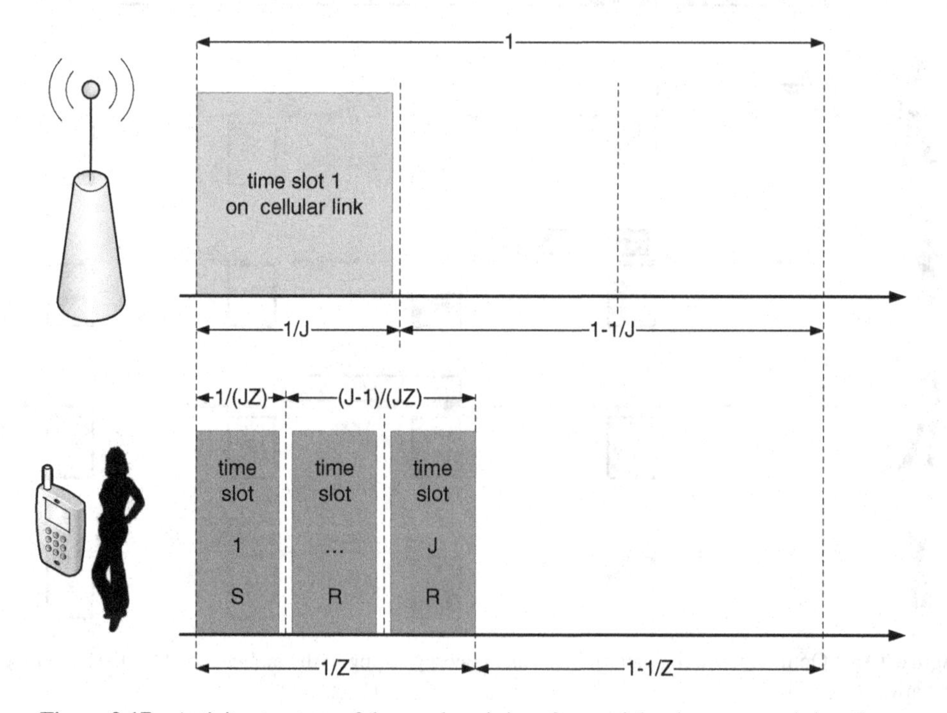

Figure 9.17 Activity structure of the overlay air interface and the short–range air interface.

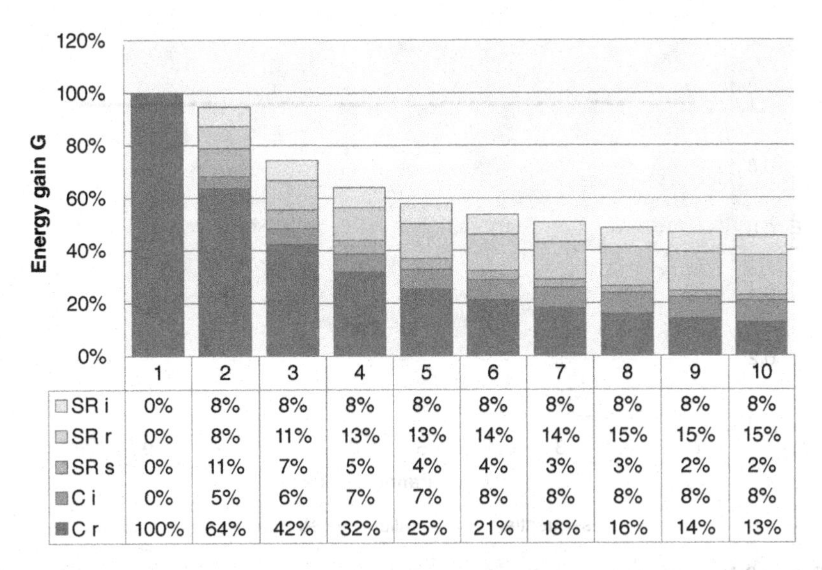

	1	2	3	4	5	6	7	8	9	10
SR i	0%	8%	8%	8%	8%	8%	8%	8%	8%	8%
SR r	0%	8%	11%	13%	13%	14%	14%	15%	15%	15%
SR s	0%	11%	7%	5%	4%	4%	3%	3%	2%	2%
C i	0%	5%	6%	7%	7%	8%	8%	8%	8%	8%
C r	100%	64%	42%	32%	25%	21%	18%	16%	14%	13%

Figure 9.18 Energy consumption for the streaming scenario versus number of cooperating mobile devices separated into energy for the cellular receiving, cellular idle, short–range sending, short–range receiving, and short–range idle (from the bottom to the top).

9.4 Comparison of the Different Approaches

In this section the three approaches are compared with each other. Figures 9.19 and 9.20 show the energy gain and the download time for the sequential and parallel cooperative download, as well as for the streaming approach. Concerning energy gain we see that the performances of the two cooperative download approaches do not differ too much.

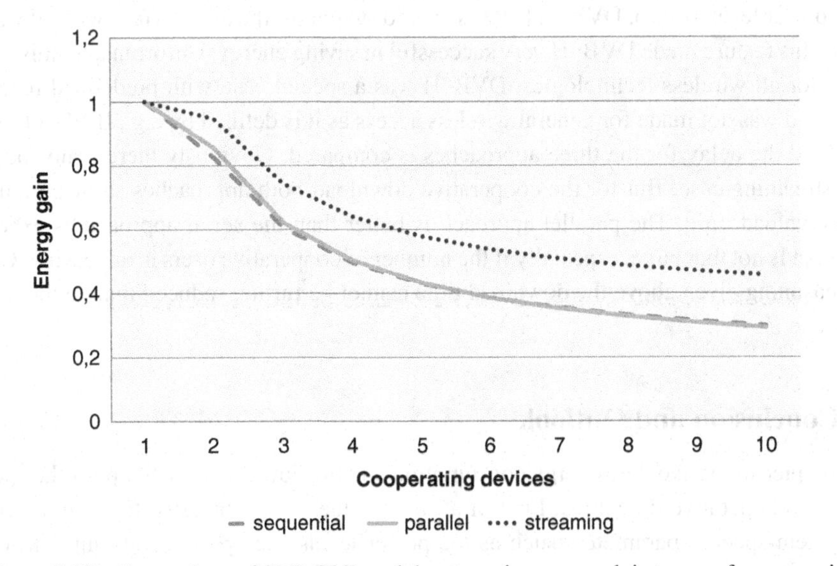

Figure 9.19 Comparison of SLE, PLE, and the streaming approach in terms of energy gain.

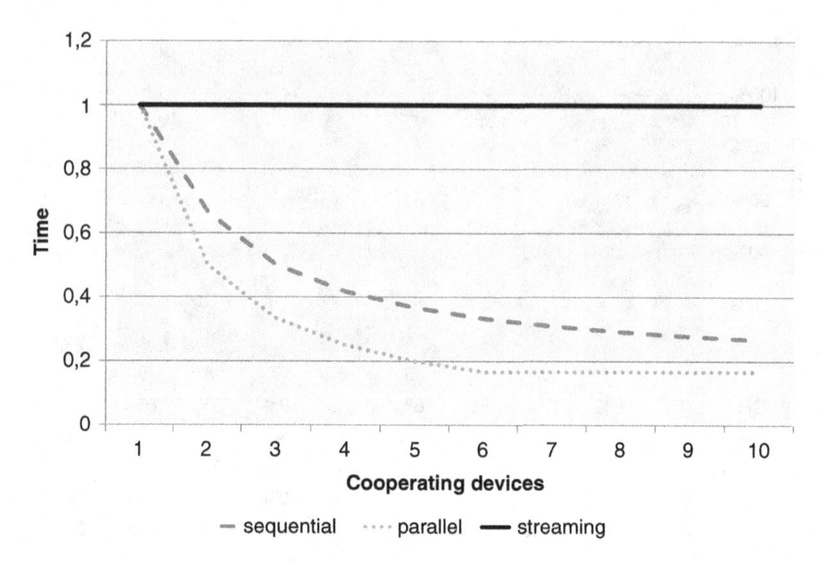

Figure 9.20 Comparison of SLE, PLE, and the streaming approach in terms of time.

It is interesting to note the important role of the power consumed by the air interfaces while being in the idle state. If idle power consumption would be zero, then the energy consumption for all three approaches would be the same. Therefore, the idle power level has the most impact on the energy consumption for cooperative communication in and with the mobile cloud. Therefore, manufacturers of wireless chips are not just focused on the data rate achievable over the air interface but also on the power level while being idle. In the early days of the Wi–Fi chip sets for mobile phones the operational time for mobile phones using Voice over Internet Protocol (VoIP) services over Wi–Fi was greatly reduced to some hours as the Wi–Fi chipsets exhibited very high power energy consumption in idle mode. In order to reduce the idle power level to zero, DVB–H [7] was able to switch off the air interface when it was not needed. This feature made DVB–H very successful in saving energy. Unfortunately this cannot be done for all wireless technologies. DVB–H was a special case with predefined receiving patterns and was not made for general wireless access as it is defined by e.g., IEEE 802.11. In Figure 9.20 the delay for the three approaches is compared. Obviously there is no speed-up for the streaming case. But for the cooperative download both approaches show a reduction in the download time. The parallel approach is better than the serial approach as expected, but the gain is not that large, especially if the number of cooperative users is increasing. Owing to the reasoning given above the download time cannot be further reduced than to the inverse of the value Z.

9.5 Conclusion and Outlook

In this chapter we derived the energy consumption and the download speed up for the specific case of the cooperative download. First of all both values under investigation are dependent on a) system-specific parameters such as the power levels (i.e., power consumption) of the technologies used, b) data rates achievable for both technologies defining the value Z) and c) scenario-dependent parameters such as the number of cooperating devices.

So far we assumed no packet losses for the wireless communication either between the user and the overlay network or between users in a mobile cloud. But in reality the communication from the overlay network towards the mobile device as well as the communication within the mobile cloud is error–prone. The error–prone link between the mobile device and the overlay network would only increase the download time linearly and losses are expected to be low compared with the errors that may occur in the mobile cloud. In [8] we have shown that Device–to–Device communication with Wi–Fi ad hoc may experience loss probabilities of up to 30%. But the communication within the mobile cloud is not only Device–to–Device, it also broadcasts information as soon as there are more than two devices participating in the mobile cloud activities. A second assumption we used before is that all mobile devices are directly connected to each other. But if we change the scenario in Figure 9.5 towards a new scenario given in Figure 9.21, we see that the exchange becomes already more complicated as one of the three partners' mobile device is only reachable via an additional hop. This extra hop will penalize the energy-saving potential. The energy saving are even more reduced if error-recovery mechanisms are considered. Therefore, it is very important that these extra costs are minimized. In general there is no way to improve the network topology within a mobile cloud.

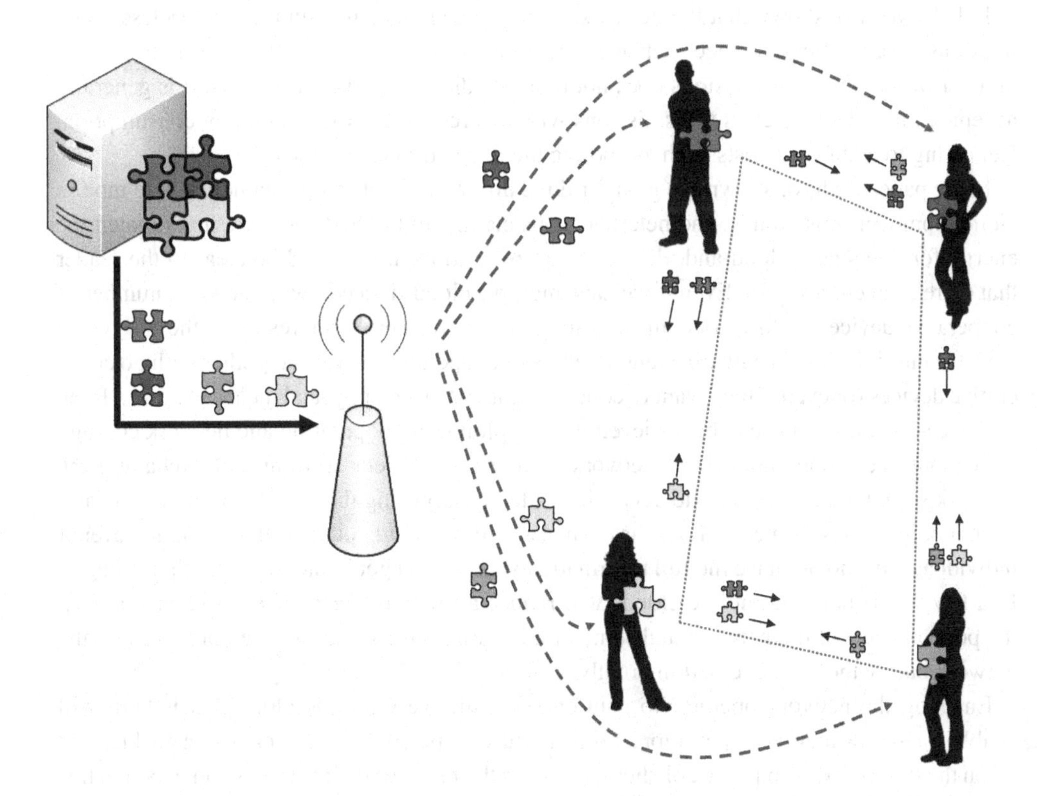

Figure 9.21 Local exchange of information in a partially meshed network.

Some attempts to introduce a degree of freedom in changing the network topology have been presented in [9] using beamforming strategies at the mobile devices, but those ideas are far from being practically realized in mobile devices and we just mention them here for the sake of completeness. An interesting way to decrease the amount of information that is exchanged within the mobile cloud, if the users are not directly connected, is to look into different seeding strategies. So far we assigned each mobile device the same amount of data. Based on the given network topology it might be beneficial to seed more information at beneficially located nodes. Those strategies were discussed in [10] and we refer the interested reader to those research results.

The most powerful improvements towards the error recovery is the use of network coding. As we have seen in Chapter 5 it will improve the communication between the overlay network and the mobile cloud as well as improving the communication within the mobile cloud.

9.6 Energy Gain for the Network Operator

So far we have investigated the energy gain for the mobile device. After reading explanations beforehand it should be clear by now why there is an energy gain for the mobile device. But the real energy costs are at the network operator and the question arises whether even the network operator can gain.

In [11] we have shown that the network operator can reduce the number of wireless transmissions for a multicast service such as Internet Protocol Television (IPTV). Even though the number of wireless transmissions does not translate directly into energy costs, it is generally accepted that fewer wireless transmissions will also reduce the overall energy consumption. But owing to additional costs such as cooling the translation is not linear.

In the paper we have shown the gain for three different cloud setups, namely single mobile cloud, homogeneous clouds, and heterogeneous clouds. In the first setup we investigated the energy for one single cloud and, due to the aforementioned, it should be clear to the reader that there is an energy gain. Even if we have multiple clouds in a cell with the same number of cooperative devices, it does not come as a surprise to save energy. Interesting is the case when we have multiple clouds with different numbers of cooperating devices together with no cooperative devices (or even if they want to collaborate they have no physical means to do so). Even in this case energy gains can be achieved if we exploit user cooperation and network coding.

Without user cooperation (and network coding for efficient seeding and exchange) the network would treat every mobile device individually, adjusting the energy for the worst case. With mobile clouds the network just has to make sure the cloud got the full information even if individual users do not have the full information yet (but will get it later over the short–range). In a few words building mobile clouds will reduce artificially the number of receivers from the point of view of the network and some of the *crying baby* situations are not solved by the network but by the mobile cloud internally.

Enabling the network operator to gain energy with every mobile cloud that is built will enable a smooth market penetration. From the users' perspective the energy gain kicks in when the first cooperation is established. And even the network operator sees some gain in this moment. The interested reader is referred to [11] for more details and the exact energy gains.

9.7 Conclusion

In this chapter we have shown the energy-saving potential for the mobile devices for different use cases of the mobile cloud. The use cases were targeting the cooperative download of multimedia content, both non-real–time and real–time services. In general we have seen that the energy-saving potential is significant for the mobile cloud compared with state-of-the-art solutions. Reducing the energy consumption is already feasible with a low number of users in a mobile cloud. A larger number of users does not lead to additional gains in the same order, but might be counterproductive as it requires more signaling among the peers. The proposed mechanism will also work if members of the mobile cloud are not directly connected but in a mesh-like topology. In this case network coding will help to keep the energy consumption for local exchange low. Furthermore we have briefly discussed the potential for network operators to preserve energy in the case where mobile cloud communication is enabled in the cell. An important fact is that already with the first cooperating devices the network operator will save energy even in the presence of devices that are not able to or willingly decline to cooperate.

References

[1] F.H.P. Fitzek and M. Katz, editors. *Cooperation in Wireless Networks: Principles and Applications – Real Egoistic Behavior is to Cooperate!* ISBN 1-4020-4710-X. Springer, April 2006.

[2] F.H.P. Fitzek and M. Katz, editors. *Cognitive Wireless Networks: Concepts, Methodologies and Visions Inspiring the Age of Enlightenment of Wireless Communications.* ISBN 978-1-4020-5978-0. Springer, July 2007.

[3] Aalborg University Mobile Device Group. Energy measurements for mobile phones. http://mobiledevices.kom.aau.dk/research/energy_measurements_on_mobile_ phones/.

[4] G.P. Perrucci, F.H.P. Fitzek, G. Sasso, W. Kellerer and J. Widmer. On the Impact of 2G and 3G Network Usage for Mobile Phones' Battery Life. In *European Wireless 2009*, Aalborg, Denmark, May 2009.

[5] G.P. Perrucci, F.H.P. Fitzek and M.V. Petersen. *Heterogeneous Wireless Access Networks: Architectures and Protocols – Energy Saving Aspects for Mobile Device Exploiting Heterogeneous Wireless Networks*, ISBN 978-0-387-09776-3 10, pages 277–304. Springer, 2008.

[6] L. Militano, F.H.P. Fitzek, A. Iera and A. Molinaro. On the Beneficial Effects of Cooperative Wireless Peer to Peer Networking. In *Tyrrhenian International Workshop on Digital Communications 2007 (TIWDC 2007)*, Ischia Island, Naples, Italy, September 2007.

[7] E. De Diego Balaguer, F.H.P. Fitzek, O. Olsen and M. Gade. Performance Evaluation of Power Saving Strategies for DVB–H Services using adaptive MPE–FEC Decoding. In *16th International Symposium on Personal Indoor and Mobile Radio Communications (PIMRC 2005)*, Berlin, Germany, September 2005.

[8] J. Heide, M. Pedersen, F.H.P. Fitzek, T. Madsen and T. Larsen. Know Your Neighbour: Packet Loss Correlation in IEEE 802.11b/g Multicast. In *4th International Mobile Multimedia Communications Conference (MobiMedia 2008)*, Oulu, Finland, July 2008. ICTS/ACM.

[9] Chenguang Lu, F.H.P. Fitzek, P.C.F. Eggers, O.K. Jensen, G.F. Pedersen and T. Larsen. Terminal-Embedded Beamforming for Wireless Local Area Networks. *IEEE Wireless Communications*, 2007.

[10] L. Militano, F.H.P. Fitzek, A. Iera and A. Molinaro. Data Seeding in Nomadic Cooperative Groups. In *Sixth Workshop on multiMedia Applications over Wireless Networks (MediaWiN) in association with the Sixteenth IEEE Symposium on Computers and Communications (ISCC 2011), Kerkyra, Greece*, 2011.

[11] J. Heide, F.H.P. Fitzek, M.V. Pedersen and M. Katz. Green Mobile Clouds: Network Coding and User Cooperation for Improved Energy Efficiency. In *IEEE International Conference on Cloud Networking (CLOUDNET)*, Paris, France, 2012.

[12] P. Karunakaran, H. Bagheri, M. Katz. Energy Efficient Multicast Data Delivery Using Cooperative Mobile Clouds, European Wireless Conference, Poznan, Poland, 2012.

Part Five

Application of Mobile Clouds

10

Mobile Clouds Applications

We are all agreed that our theory is crazy. The question which divides us is whether it is crazy enough to have a chance of being correct. My own feeling is that it is not crazy enough.

Niels Bohr

In this chapter we discuss several applications that exploit or support the mobile cloud concept. Some applications are already commercially available, while other applications described in this chapter are still in the lab phase. The applications are grouped by their architecture type and the cooperation form explained in previous chapters.

10.1 Introduction

The cloud concept is very popular these days and even mobile clouds are getting more attention lately. Mobile clouds are entering the market place of mobile apps. Some readers might even have used one or more commercial realization forms of the mobile cloud, maybe under a different name such as mashup or swarm intelligence. Nearly all of these realizations come in the form of a mobile app that needs to be installed by the user, and less frequent are the realizations done as web services. For the user it does not make a big difference whether to use a native app or a web service. Without going into detail, the web services are easier to deploy as they are less platform dependent, but native apps offer more flexibility in programming as well as additional access to programming interfaces referred to as application programming interfaces (APIs). The latter are needed to get access to hardware sensors or short-range communications technologies. With the further development of the HTML standard this might

Mobile Clouds: Exploiting Distributed Resources in Wireless, Mobile and Social Networks, First Edition.
Frank H.P. Fitzek and Marcos D. Katz.

change, but with the current version HTML5 there are still limitations when it comes to some examples of mobile clouds. Most mobile apps require other users to also install the app in order to generate an additional value for the users as usability increases with each new user joining the mobile cloud. It is worth noticing that some cooperative approaches do not rely on additionally installed mobile apps though mobile users can still contribute to the mobile cloud without even knowing it.

In this chapter we will describe different realizations in detail and categorize them by their architecture and the cooperation form. With reference to Chapter 2 there are two main architectures for mobile clouds. In the first architecture mobile users are connected via an overlay network with each other and therefore are referred to as Overlay Mobile Cloud (OMC). In order to organize those mobile devices connected to the same service there might be a dedicated cloud service managing the devices and their resources. The second architecture is not that common as it also includes a possible direct connection among mobile devices in close proximity and is referred to as Short–Range Mobile Cloud (SRMC).

The cooperation forms are described in Chapter 8 and we just list them here for convenience: namely forced cooperation, technically enabled cooperation, socially enabled cooperation, and altruism. Combining those four cooperation forms with two architectures leads to eight different solution spaces. We note that the categorization is not always that clear and some use case may overlap to more than one cooperation field, but then we will list it where we believe there is a dominant field of cooperation.

In Table 10.1 several possible architecture types and cooperation forms are listed together with a short explanation and some examples. In the following we describe those different realization forms of mobile clouds. In Sections 10.2 to 10.5 we will describe those cases where the communication architecture relies on the overlay network solely. Note that those examples are sometimes related to the concept of swarm intelligence. The differentiation in this chapter lies in the cooperation motivation for the individual swarm members. In Sections 10.6 to 10.9 the communication architecture becomes more evolved, including also direct communication among mobile devices. In Sections 10.2 to 10.9 the four different cooperation cases are described.

10.2 Forced Cooperation – Overlay Network

In this subsection we focus on mobile clouds, where the mobile users are solely connected to an overlay network. Even though the mobile user might not necessarily install a mobile app, nor intend to be connected nor cooperating with anybody, there are situations where crowd–sourced information can be exploited by others. It was always suspected that the mobile devices secretly collect information without the user being aware of it. But only lately has irrefutable evidence in the press shown that this really happens. Furthermore, some network operators plan to sell anonymous data of their customers to third parties. Readers might question the legality of this approach, but we are just labeling it forced cooperation without discussing the ethical justification of it. Three examples are presented next, the first one related to a network operator, the second one to mobile device manufacturers and the last one to mobile app developers.

Table 10.1 Mobile application or demonstration of different mobile cloud realizations.

MC type	Coop. type	Operating principle	Example
Overlay MC	Forced	Network operators, mobile manufactures and mobile apps are monitoring user behavior in order to tailor their services. Users are often not aware about their contribution to the cooperation and have nearly no possibility to deny the service. E.g., background applications gathering information of the mobile user such as cell ID, GPS position, and SSIDs of Wi–Fi hotspots.	Telefonica, Apple, Google
	Technical	Users install an extra mobile app on their mobile phone connecting to an overlay cloud service. There is a clear advantage for the user to use the tool and also to contribute to cooperation. Here the users are aware of the data that is shared among the user community and happy to do so as there is a clear gain.	Waze, TomTom
	Social	Users are contributing to a cloud service to get known in the comment, get rewarded by comments or virtual points and improve e–reputation.	[1, 2]
	Altruism	Users do not care about the gain and contribute altruistically to cloud services.	[3, 4]
Short–Range MC	Forced	Mobile devices are forced into cooperation in order to achieve a system gain or satisfy the user.	Tethering your device, users collaboratively use their devices.
	Technical	Users share their resources with others and gain is visible instantaneously.	CoopLoc
	Social	Users share their own resources with others and gain social respect and reputation for this interaction.	Gedda–Headz, InstaBridge, Open Garden
	Altruism	Sharing resources of one device with others without retrieving anything back. One of the first examples was the Joiku [5] mobile app sharing an IP connection of an overlay network with locally connected devices.	Joiku

Different cases are grouped by their architecture type and cooperation approach approach.

10.2.1 Crowd–sourced Information by the Network Operators

Very recently [6] it was announced that the network operator Telefonica will offer 3rd parties access to analytics on anonymous location data. Other network operators have announced similar plans for the near future, not only offering location data but also cars' speed information.

Owing to the network architecture and protocols, network operators have been always able to locate mobile phones in order, e.g., to support police forces to track down suspects or to identify the location of stolen phones. In [6] data is planned to be made available to third parties in order to provide information about user mobility. There are many ways to exploit this set of data, for instance, for effective advertisement placement across the city. But also public transportation systems could be optimized in case user location is known. If a large number of users are in certain area, more means of transportation could be allocated to that area region, and conversely, the number or public transportation vehicles could be reduced in those areas with fewer people at a given time. In this case the cooperation as such is not visible to users but they do contribute to creating a valuable service – whether they want or not.

10.2.2 Crowd–sourced Information by the Manufacturers

Other recent initiatives [7, 8] describe plans by which mobile phone platforms will be able to track location data and other user-sensitive information as a basic feature of the mobile manufacturer's firmware. While the marketing idea is to make users' life easier, the underlying goal of manufacturers appears to be to increase their power over their competitors. By logging location with contextual information e.g., which Wi–Fi hot–spots are available at that given location, the manufactures are able to get a very precise overview of a Wi–Fi coverage map. This approach is especially interesting if the same manufactures have other tablets or readers that only are equipped with Wi–Fi technology. Those devices can then retrieve the location just based on the service set identifier (SSID) of the Wi–Fi hotspot without the need of cell ID or GPS information. This is in principle rather harmless, and even appears as a nice way to help each other. But on the other hand, the contextual data that is stored on a device contains also the Wi–Fi login credentials. As this happens without the consent of the mobile user we refer to it again as forced cooperation.

10.2.3 Crowd–sourced Information by the Mobile Apps

The last example refers to mobile apps as data collectors. In the past [9] there have been several cases where mobile apps secretly uploaded private data to a centralized cloud storage. By doing that developers will be able to know more about the user and to combine the information from million of users. The new currency is *Big Data* and the number one in the space in undoubtedly Facebook. But also other companies such as LinkedIn, Twitter and Path are playing similar games. It is clear that those companies will underline the power of *Big Data* to make our lives easier and that by our cooperation we all contribute to this success whether we like it or not. Users rarely complain about the stolen data problem as in the cases of Path and other companies, that secretly looked into the phone book of their users without informing them about it. The question that may arise next is why social networks are interested in phone book data anyway as they have the social graph already? The reason is that the real social graph of a user in not determined in the social network itself but is hidden in the phone book. Facebook and others are just the second tier friendship; the real connections are in the phone book.

10.3 Technology–enabled Cooperation – Overlay Network

In this section we assume that mobile users are willing to cooperate as they receive a benefit for their cooperative contribution and they are not forced to contribute as in the previous subsection. There are several examples but we limit ourselves to [1, 10, 11]. Owing to the widespread distribution of mobile phones, environmental information could become crowd–sourced. This means that mobile devices-generated sensor data is forwarded to a central cloud service for the use of authorities and other mobile users. This approach is much more cost-effective than deploying dedicated sensors across large areas (e.g., city-wide). For example, the mobile app Waze [11] uses the GPS data of a mobile phone to understand the location of a vehicle. The user sees a clear benefit in the mobile app as it offers map material and the option to route to a certain target as shown in Figure 10.1.

The information gathered (location, speed, destination) is shared among other users via a cloud solution in order to estimate traffic conditions around the user. This means if a user experiences a traffic jam it will warn other users not to take that route. All other users will benefit by this information. From a cooperation point of view this approach is interesting as the pay–off for the user in the traffic jam is delayed. This means the user will receive a gain out of her/his cooperation at a later point in time, for example when she or he is warned about certain traffic situations. Crowd-sourced traffic management has some advantages as it happens in real time. The disadvantage is clear, if the density of contributing vehicles is low (traffic information may become obsolete anyway). Therefore, this mobile app is based on the assumption that as many drivers as possible participate in the service. In addition to traffic data, other information such as speed cameras [1, 2, 11] and fuel prices [11] are exchanged among users. Moreover, the route of each user can help to maintain the map material, thus keeping it up to date. In [10] TomTom adapted that idea to their own hardware and software platform.

Currently crowd-sourced approaches collect as much sensor data as possible. However this approach aiming for *Big Data* has some drawbacks. Although additional users are desired by these systems, if a large number of sensor nodes are reporting back to and receiving processed data from the cloud, the system will face challenges at three levels: (i) network: due to limitations in the bandwidth on the wireless links, (ii) mobile device: due to constraints on energy consumption related to pre–processing of the data, and the rate of sensing and reporting to the cloud, and (iii) cloud: due to the computational complexity needed to process and exchange information on the time-varying state of the system based on sensors measurements. The main issue and also opportunity is that the information of different mobile nodes will be redundant due to correlations in the underlying processes that are being measured. By exploiting user cooperation and network coding, the aforementioned problems can be reduced. The main idea is to identify sensor data that is locally correlated and process that information (reducing the number of samples and deriving meaningful statistics). This pre-processed data will be network coded and sent by a subset of mobile devices in order to ensure a reliable transmission to the cloud.

It is interesting to note that as of today most high-end cars still use stored data on DVDs for the map material and overlay information from the traffic message channel (TMC). The

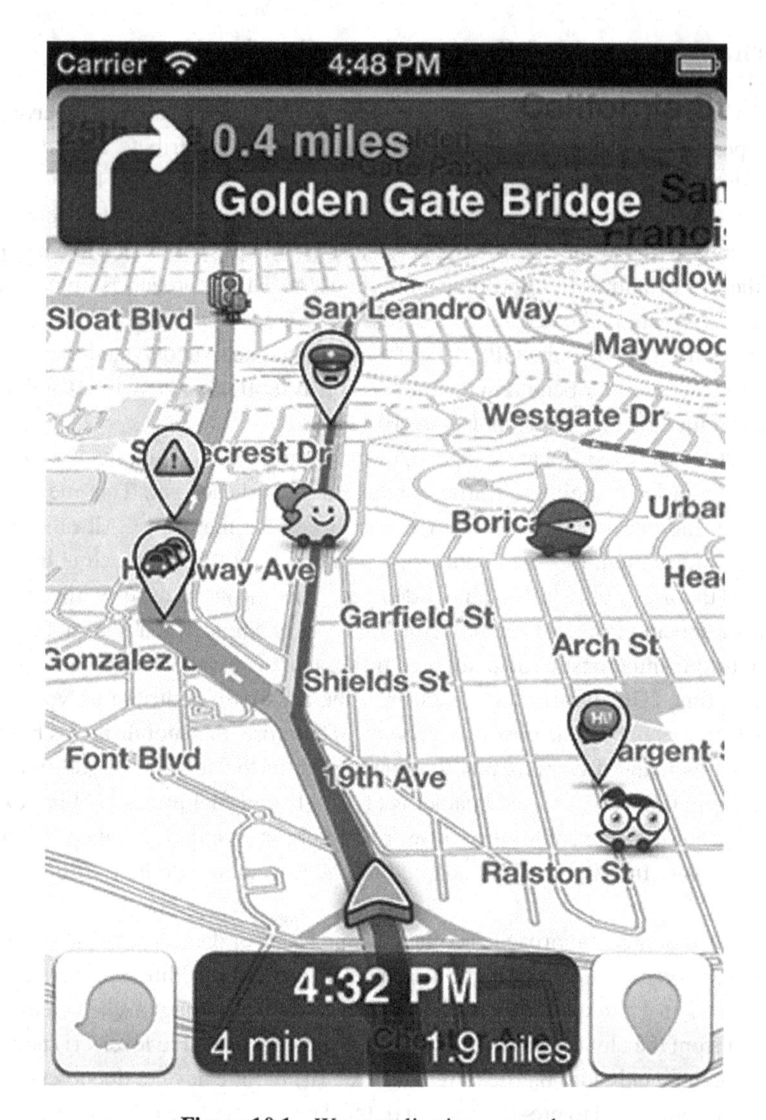

Figure 10.1 Waze application screenshot.

information of the TMC is gathered from external sensors. But this approach, compared with the crowd–sourced approach in [1, 10, 11], is not real time and therefore often delayed. One can say that one of the key reasons crowd–sourced information is not widely used today is that car manufactures are concerned that their customers may feel that they are being monitored.

10.4 Socially–enabled Cooperation Overlay Network

Most cloud-based application have a *social touch.* All applications use social networks to promote and market themselves to a larger user base and therefore all applications have already some social element. In our previous example considering users in cars warning each other about speed cameras [1, 2] the social element plays a larger role. Even though it is technically enabled, the social domain is important. Each warning that is added to the system will be labeled with the user name and other users can thank this user for the given warning (regardless of whether they needed it or not). In applications such as [10, 11] there is no direct link between the data added to the system and the individual user as it is for the speed warnings.

10.5 Altruism – Overlay Network

In this section we discuss mobile apps that contribute to a cooperative service based on altruism. In [4] the service QYPE allows users to judge on locations, events, food, and many others. Each user votes for the service they get. In [3] the company barcoo introduced a interesting application. Users contribute to a centralized service by uploading nutrition information about certain products without expecting any reward for their work. For the users there is no gain in doing it but they are driven by the personal satisfaction of anonymously helping others, as well as their belief that even small contributions help to build a better world. Therefore, here we refer to the action as being altruistic. As we stated before the characteristics between the different cooperation fields are not always clear and it could be argued that users that contribute to such a system trust in the fact that others will reciprocally provide the information that might be useful for them in the future.

10.6 Forced Cooperation – Direct Mobile Cloud

Now we consider mobile clouds that are made up of mobile devices in close proximity using the short–range links to communicate with each other in addition to existing overlay communication links. In case one user has several mobile devices, the user might want to use those devices in a collaborative way instead of each device independently. The gain will not be optimized for a single device but for the user. Therefore, the cooperation among devices can be seen as forced. For example, a user may use his or her own mobile device to connect a laptop to the Internet. The laptop can be connected to the mobile device via Bluetooth or Wi–Fi while the mobile device uses the cellular connection to connect to the Internet. Here the cellular resources of the mobile device are shared with the laptop. While the mobile device invests energy to support the laptop, the gain for the laptop is clear. But the cooperation will be established as the user owns both devices. Note that this case of forced cooperation can also be considered as self–cooperation, where the devices establish cooperation to benefit their owner.

Another very interesting solution for aggregating resources has been presented in [12] where multiple screens of different mobile devices are pooled together. The screens can belong to one user or to several users. As shown in Figure 10.2 the devices are just placed randomly or with the aim to create an optimal screen (most likely a rectangle with a ratio close to 4 : 3 or 16 : 9). Afterwards the sending device (not among those forming the screen) will ask each mobile phone to show a QR code as illustrated in Figure 10.2. The QR code helps to figure out the

Figure 10.2 Positioning of multiple mobile devices with different form factors to collaboratively building one large screen showing QR codes to help detect positions and orientations.

Figure 10.3 Final outcome of the collaborative screen.

Figure 10.4 16 iPods are receiving content from one source and playing it synchronously [13].

orientation of the mobile phone as well as it conveys additional information about the screen size and the way it can be connected. The sending device will then make a picture of all QR codes and send the proper component parts of the picture, according to the position and size of the QR code, to the mobile devices. Figure 10.3 shows the composed puzzle–like picture by multiple phones. Note that this example seems to be more organized than just randomly placing the mobile devices on the ground, but users will tend to create a rectangular form.

At this point we would like to point to Chapter 3 for more examples. Besides combining the displays as explained here we have already pointed out the example of sharing loudspeakers (Section 3.3), sharing microphones (Section 3.4), or sharing camera functionalities (Section 3.5). In all of these examples the devices are forced to cooperate by their users in order to get a better performance. We also would like to underline that these use cases are not just concepts but have been implemented at Aalborg University. For instance the collaborative use of speakers has been realized for multiple iPods. In [13, 14] one device streams video and audio to 16 iPods that played the content synchronously over an equal number of screens and loudspeakers. Figure 10.4 shows the setup used for this experiment whereas a video of the implementation is available at [15].

10.7 Technically–enabled Cooperation – Direct Mobile Cloud

Here we describe mobile clouds that will offspring because each user will see an instantaneous gain. We list several examples here as this form of mobile cloud is the most promising one.

10.7.1 CoopLoc

A very interesting example of cooperative location was presented in 2008 by Sammarco et al. in [16]. The main idea was to exchange rough localization estimates among mobile phones in order to achieve a more accurate estimation for each individual device. Note that at that time GPS localization was just giving its initial steps in the mobile phone ecosystem. But GPS worked only outdoors and consumed significant amounts of energy, draining the battery of the mobile phone and hence making its use not appealing. The idea of cooperative localization was to exchange the localization estimates by Cell IDs among mobile phones. A single mobile device can retrieve the cell ID from the base station it is connected to, and derive a rough estimate of its current location. The advantage is that this works also inside and consumes less energy compared with the GPS approach. In this context, Google launched an application at the same time called My Location that exploited single cell IDs to show mobile users approximately where they are located on Google maps. The downside is that the estimate was rather inaccurate compared with GPS. Google stated that the accuracy of My Location was in the order of 1000m. Network operators have the chance to use triangulation so exploiting the information from multiple base stations of their own. This approach is very accurate but only accessible for the mobile network operators. Some operators offer this approach as a service to retrieve the localization of a mobile phone.

From a developer's point of view there is only one cell ID available, even though the mobile phone has information on all base stations that are surrounding the device. The mobile device has to know all available base stations in order to prepare for potential hand overs. A *quick hack* from a Russian programmer had shown how to get the list of all base stations for a Nokia 6600 device. Having the full list of cell IDs would allow the developer to derive a better localization as shown in Figure 10.5. But this approach was rather complex and not applicable to other devices.

In order to get multiple cell IDs even though only one cell ID is available per mobile device, cooperative localization exchanges localization information among mobile device in close proximity, producing then a more accurate estimate by combining those estimates into one. As mentioned before only one cell ID can be retrieved per device using the programming APIs of the mobile device. The local exchange was realized by Bluetooth, but could be done with other technologies, such as Wi–Fi. However, Bluetooth had several advantages. First of all, the devices had to be close to each other in order to exchange their estimates and therefore they are approximately in the same location. Wi–Fi has a larger coverage than Bluetooth and is therefore less suited. Furthermore, Bluetooth comes with its own service discovery protocol, which together with the Bluetooth technology results in a low-energy solution. The only disadvantage is that Bluetooth has no broadcast mode to exchange localization estimates.

In order to validate the approach a full measurement campaign was carried out in Aalborg, Denmark. Initially, three mobile phones with SIM cards of different network operators, namely CBB, TDC and Telia, were used to retrieve the cell IDs for every street in downtown Aalborg (see Figure 10.6). In order to retrieve the cell ID a small Python script for Symbian device was written. The script logged the location with GPS and retrieved the cell ID. The measurement campaign was repeated at different days and at different times of the day. The density of

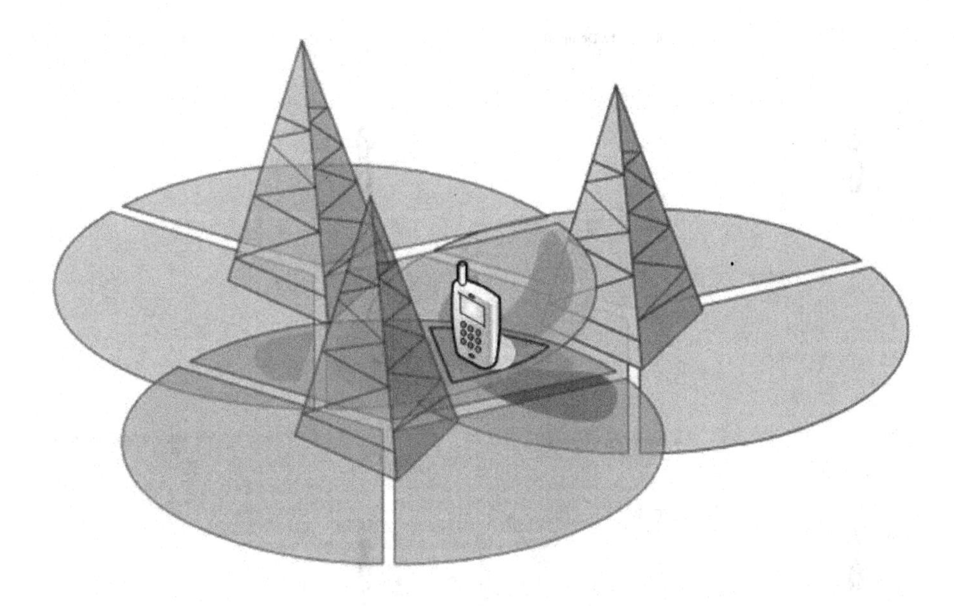

Figure 10.5 Example of cooperative location: CoopLoc.

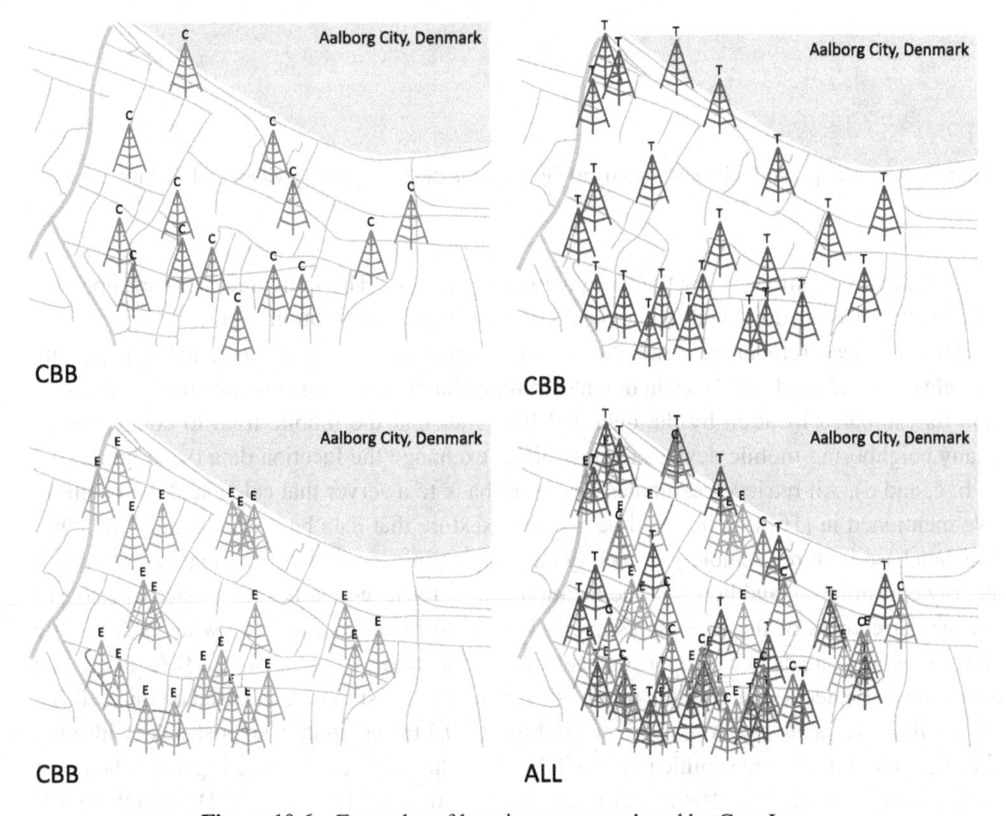

Figure 10.6 Examples of location maps produced by CoopLoc.

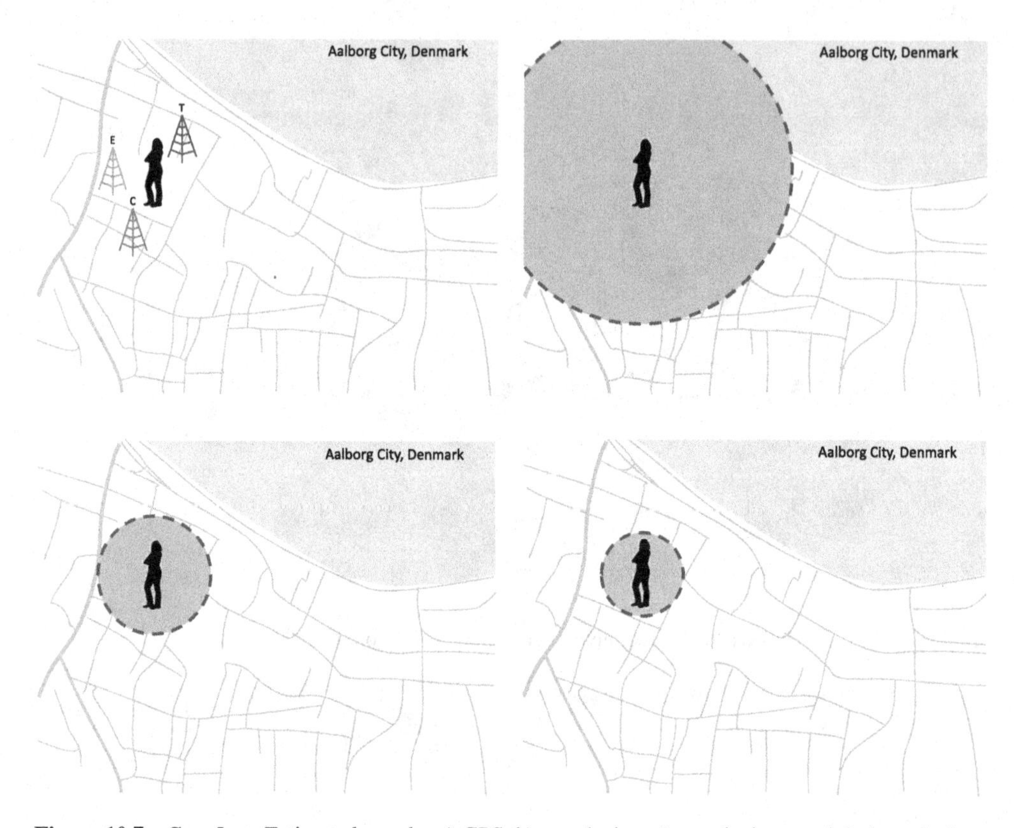

Figure 10.7 CoopLoc: Estimated area by a) GPS, b) one device, c) two devices, and d) three devices.

base stations was different for the three network operators. The individual network operator information was accumulated in one data base (see Figure 10.6 bottom right).

After the measurement campaign the cooperative location approach was tested by using one mobile device at random places in downtown reporting its GPS position, just for comparison, and its estimated location by the own cell ID. After that the mobile tried to connect to as many neighboring mobile devices as possible to exchange the location data (see Figure 10.7 a, b, c, and d). All retrieved estimates were sent back to a server that calculated the position. We mentioned in [16] that it would be possible to store that data base even locally. Whether the data base is stored remotely or locally has only an impact on the performance in terms of energy consumption and how fast the location can be retrieved but has no impact whatsoever on the correctness of the location estimate. Figure 10.7a shows the setup with three mobile devices, where one device is reporting back the GPS location. Figures 10.7b, 10.7c, and 10.7d show the estimated location of the mobile device when the estimation is based on one cell ID, two or three, respectively. As can be seen in Figure 10.7 the estimation error shrinks with each additional mobile device significantly. In Table 10.2 the uncertainty range is given. The mean value of the uncertainty is 168m, 121m, and 74m for one cell ID, two cell IDs, and three cell IDs, respectively.

Table 10.2 Estimation error of cooperative localization.

Number of devices	Minimum Distance	Mean Distance	Maximum Distance
1	14.89m	168.62m	668.86m
2	3.80m	121.21m	660.24m
3	5.29m	74.46m	243.64m

The given values correspond to the area where the Cooploc approach assumes the users are in, compared with a location retrieved with GPS. Therefore, the smaller the area the better the estimate.

At this point we have to state that the setup favors our scenario as the cooperating mobile devices always have different network operators. In real scenarios it might be that a mobile phone will ask for location estimates of mobile phones that have the same network operator and therefore potentially the same cell ID, which would not help to increase the accuracy. Also the trend of cell tower sharing will have a negative impact on this approach. Nevertheless the aim of this project was to show the power of cooperation by simply exchanging small data within the mobile cloud and improving an existing service significantly.

10.7.2 Cooperative Access

The cooperative access case was already mentioned in previous chapters. Here mobile users share their overlay access in a cooperative manner in order to consume cooperative services. In Figure 9.2 the main topology is shown. To the best of our knowledge no such service using cooperative access is commercially available yet. But we have shown an implementation for cooperative download [17] and cooperative web browsing [18]. The main difference between the two use cases is that the cooperative download provides instantaneous gain for all users, while the cooperative web browsing requires some pay–off tolerance in the order of some seconds. In contrast to Section 10.8.1, cooperative access is not just relaying the communication over another mobile device, but all participating devices are interested in using the services at the same time such as video streaming of sport events. In other words, each device will gain immediately (pay–off tolerance up to some seconds).

10.8 Socially–enabled Cooperation – Direct Mobile Cloud

10.8.1 Sharing Internet Connections

In Chapter 8 we have already discussed an example for sharing Internet connection among users. In Figure 8.3 it is shown how one user shares the Internet connection with another user. Social elements such as posting on Facebook are exploited to thank the user for sharing the Internet connection. The first attempts at exploiting ideas on sharing Internet connections are discussed in [19, 20].

10.8.2 Sharing Applications

Concepts and implementation initiatives related to sharing applications are not well developed yet, still there is already some activity in this promising area. Here we refer to our own work implementing a mobile game called Gedda–Headz [21]. The game could be played between two players, wirelessly via Bluetooth or via the Internet connection. As with all mobile apps, the question was how to virally distribute the content to users that are not even aware of it as they may not have Internet connection or only sporadic Internet connection. The game was targeting the S40 platform of Nokia and the players were heavily represented on the Asian continent. The idea was to allow already hooked players to pass the game to their friends. Therefore, the Gedda–Headz community was informed about a new mobile app called *Gedda–Headz spreader* and the users were asked to install the app in a Device–to–Device fashion. The interesting issue is how to motivate the players to do the viral distribution, as explained in [22, 23]. A few people used the spreader without any reward for doing so. But the real usage of the spreader started once the users were rewarded by the application. As most games and services nowadays, Gedda–Headz provided points that users would gain if they play games against each other. The points could then be used to purchase real goods on the related web page. Therefore, it was quite obvious to use those points in a similar fashion in the spreader and reward every installation by points. As the system was aware about who was spreading the content it could also reward the spreader for his activity. This simple example shows how cooperation can be fostered in the future. All services that nowadays still rely on altruism and just good hope that cooperation might be a good thing, will soon be rewarded by social elements such as post in social networks or application-specific rewards as in Gedda–Headz [21].

10.9 Altruism: Direct Mobile Cloud

Cooperation among mobile devices based on altruism is discussed here. One example is to share the Internet connection with friends. Nowadays there are already a large number of applications implementing such a use case. The very first player in this domain was the mobile app Joiku [5]. The idea of the app was to span a Wi–Fi mobile hotspot with your mobile phone and share your IP connection that is established via the cellular operator with your friends. In Figure 10.8 a simple example is given. Anna has an IP connection with her phone using LTE connections. As Anna's network plan allows her to use the Internet with a flat fee she is willing to share the IP connection with Laura and Frank. In order to do that Anna opens an app that is spanning the Wi–Fi mobile hotspot and eventually lets her configure the access credentials for her friends. Access credentials are needed in case Anna wants to make sure only Laura and Frank may join her hotspot. Free riders would reduce the data connection among her friends and also drain her battery even more. After Laura and Frank are connected all incoming and outgoing traffic is relayed via Anna. The relaying will consume more energy on Anna's device but she is happily sacrificing it in order to help her friends – pure altruism.

The first implementation was done for Symbian devices and still the service is offered for Nokia Symbian Anna and Belle phones. The idea was based on the fact that high–end mobile

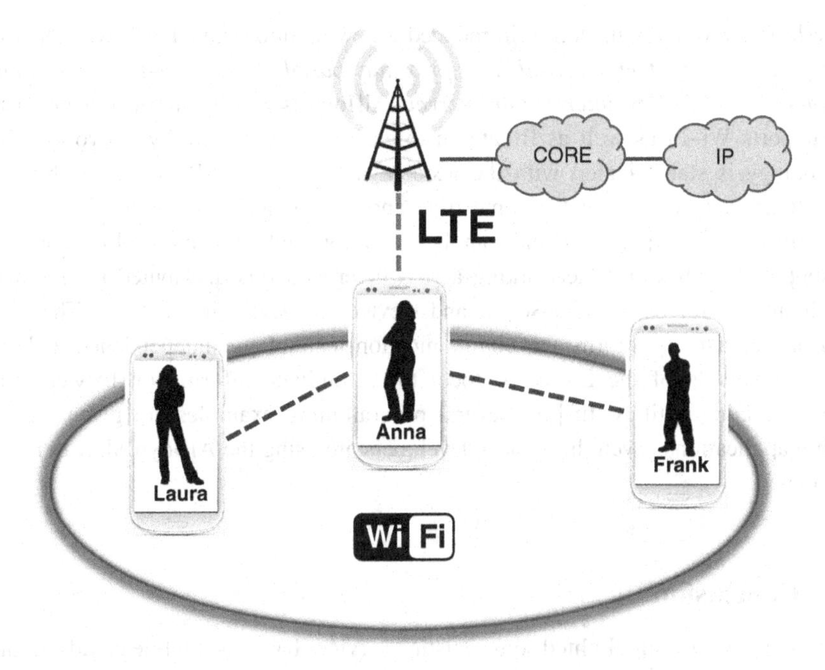

Figure 10.8 Example of altruistic Internet access sharing.

phones at that time had Wi–Fi on board and network operators offered flat–fees for mobile Internet connections. Joiku customers used the mobile app to allow their secondary devices, e.g., the laptop or even their friends' devices to connect to their service. Even though the battery of the mobile phone drained more quickly by allowing your friends to connect, altruism drove this kind of user scenario. Joiku improved their service by making the connection setup easier using Facebook or Twitter login. This shows that the connection is offered mostly to friends in order to keep the altruism idea alive. At the time this book was written Instabridge [19] and Open Garden [20] started to offer the same service for Android users. Even though the technology idea behind Instabridge is the same as that of Joiku, Instabridge focuses more on Facebook elements in order to ease promoting, and motivating cooperation among users can be seen in these examples.

10.10 Industrial Activities

There are already several attempts by the industry to enter the space of mobile clouds. Some players such as Apple and Microsoft were already and are still active in filing patents in this area. In [24] a mobile device collaboration and the exchange of resources is described. The use of a helper device to grant access is given in [25] and a Device–to–Device method of installing apps is mentioned in [26]. To all of these patents there exist prior art and we just refer to our book activities in 2006 [27] and 2007 [28]. Nevertheless, it shows that there is interest also in many key players in the field to investigate the concept of mobile clouds.

In the last few years, Qualcomm introduced a new solution called AllJoyn [29] as *peer-to-peer technology that enables ad hoc, proximity-based, Device–to–Device communication without the use of an intermediary server.* AllJoyn is a software developer kit (SDK) and it supports Wi–Fi as well as Bluetooth to easily set up proximity networks. The network topology is star oriented with the need of a local server. By doing so both Wi–Fi and Bluetooth can be used at the same time and only unicast communications are used. The solution is cross platform and can be used for mobile app development. AllJoyn offers support for proximity networking across heterogeneous distributed mobile systems, also with support for both client–server and Device–to–Device interactions. The design is focused on language, platform and communication technology independence. It leverages an extended version of the D–bus protocol to disseminate information between the distributed interacting entities. In [29] several programming examples are given for gaming and utility applications. Even third party developers are using the AllJoyn SDK to make new applications.

10.11 Conclusion

In this chapter we have highlighted some existing services based on mobile clouds. In the near future those services will evolve and new, more advanced services will come up.

References

[1] blitzer.de. web site. http://www.blitzer.de/nutzer.
[2] Trapster. Speed Trap Sharing System. http://trapster.com/.
[3] Barcoo. barcoo keeps you and your friends informed – anywhere, anytime. http://www.barcoo.com.
[4] QYPE. Qype – find it! share it! http://www.qype.co.uk/.
[5] Joiku. Joiku web page. http://www.joiku.com, 2013.
[6] BBC News Technology. Telefonica hopes 'big data' arm will revive fortunes. http://www.bbc.co.uk/news/technology-19882647, October 2012.
[7] The Guardian – News – Technology. Android phones record user-locations according to research. http://www.guardian.co.uk/technology/2011/apr/21/android-phones-record- user-locations, April 2011.
[8] The Guardian – News – Technology. iPhone keeps record of everywhere you go. http://www.guardian.co.uk/technology/2011/apr/20/iphone-tracking-prompt s-privacy-fears, April 2011.
[9] The Guardian – News – Technology – Battle for the Internet. Big Data age puts privacy in question as information becomes currency. http://www.guardian.co.uk/technology/2012/apr/22/big-data-privacy-infor mation-currency, April 2012.
[10] TomTom. TomTom HD Traffic. http://www.tomtom.com/en_gb/services/live/hd-traffic/#tab:tab3.
[11] Waze. Outsmarting traffic, together. http://www.waze.com/.
[12] R. Borovoy and B. Knep. Junkyard Jumbotron by MIT's Center for Future Civic Media. http://jumbotron.media.mit.edu/.
[13] P. Vingelmann, F.H.P. Fitzek, M.V. Pedersen, J. Heide and H. Charaf. Synchronized Multimedia Streaming on the iPhone Platform with Network Coding. *IEEE Communications Magazine - Consumer Communications and Networking Series*, June 2011.
[14] P. Vingelmann, F.H.P. Fitzek, M.V. Pedersen, J. Heide and H. Charaf. Synchronized Multimedia Streaming on the iPhone Platform with Network Coding. In *IEEE Consumer Communications and Networking Conference - Multimedia & Entertainment Networking and Services Track (CCNC)*, Las Vegas, NV, USA, January 2011.
[15] P. Vingelmann and F.H.P. Fitzek. 16 iPods. youtube, 2010.

[16] C. Sammarco, F.H.P. Fitzek, G.P. Perrucci, A. Iera and A. Molinaro. Localization Information Retrieval Exploiting Cooperation Among Mobile Devices. In *IEEE International Conference on Communications (ICC 2008) - CoCoNet Workshop*, May 2008.

[17] L. Militano, F.H.P. Fitzek, A. Iera and A. Molinaro. On the Beneficial Effects of Cooperative Wireless Peer to Peer Networking. In *Tyrrhenian International Workshop on Digital Communications 2007 (TIWDC 2007)*, Ischia Island, Naples, Italy, September 2007.

[18] G.P. Perrucci, F.H.P. Fitzek, Q. Zhang and M. Katz. Cooperative mobile web browsing. *EURASIP Journal on Wireless Communications and Networking*, 2009.

[19] Instabridge. Instabridge web page. http://www.instabridge.com, 2013.

[20] Open Garden. Open Garden web page. http://opengarden.com, 2013.

[21] Gedda–Headz. Gedda–Headz web page. http://www.geddaheadz.com.

[22] C. Varga, L. Blazovics, W. Bamford, P. Zanaty and F.H.P. Fitzek. Gedda–Headz: Social Mobile Networks. In *ACM MSWiM 2010*, Bodrum, Turkey, October 2010.

[23] C. Varga, L. Blazovics, H. Charaf and F.H.P. Fitzek. *Social Networks: Computational Aspects and Mining – User cooperation, virality and gaming in a social mobile network: the Gedda-Headz concept*, chapter 23, page 1. Springer, 2011.

[24] S. Li, Y. Zhang, G.B. Shen and Y. Li. Mobile Device Collaboration. Technical Report, United States Patent Application 20080216125, September 2008. http://www.freepatentsonline.com/y2008/0216125.html.

[25] D. Low, R. Huang, P. Mishra, G. Jain, J. Gosnell and J. Bushx. Group Formation Using Anonymous Broadcast Information. Technical Report, United States Patent Application 20100070758, March 2010. http://appft.uspto.gov/netacgi/nph-Parser?Sect1=PTO1&Sect2=HITOFF&d=PG0 1&p=1&u=%2Fnetahtml%2FPTO%2Fsrchnum.html&r=1&f=G&l=50&s1=%2220100070758%2 2.PGNR.&OS=DN/20100070758&RS=DN/20100070758.

[26] E.D. Steakley. Installing applications based on a Seed Application from a Separate Device. Technical Report, United States Patent Application 12/483,164, June 2009.

[27] F.H.P. Fitzek and M. Katz, editors. *Cooperation in Wireless Networks: Principles and Applications – Real Egoistic Behavior is to Cooperate!* ISBN 1-4020-4710-X. Springer, April 2006.

[28] F.H.P. Fitzek and M. Katz, editors. *Cognitive Wireless Networks: Concepts, Methodologies and Visions Inspiring the Age of Enlightenment of Wireless Communications*. ISBN 978-1-4020-5978-0. Springer, July 2007.

[29] Qualcomm. AllJoyn. http://www.alljoyn.org/.

Part Six

Mobile Clouds: Prospects and Conclusions

11

Visions and Prospects

Sharing is good, and with digital technology, sharing is easy.

Richard Stallman

This chapter presents visions and prospects of mobile cloud technology by shedding some light on possible future development paths. The roles of the key related players in the development paths are first considered. Moreover, enabling, supporting and complementing technologies are also discussed. New scenarios as well as novel promising conceptual applications for mobile clouds are then considered. Finally this chapter proposes a long–term vision on how mobile clouds could become one of the enabling technologies for a general-purpose, global resource-sharing concept, the supporting platform of Shareconomy.

11.1 Some Insights on the Future Developments of Mobile Clouds

Several mobile cloud concepts and concrete solutions were described and discussed in this book. One may wonder how and when mobile cloud solutions will find their way to the real practical world, and who are the key players supporting the practical development and further dissemination of mobile cloud technology. Current technology already offers the basic building blocks to practically implement mobile clouds and exploit distributed resource sharing. Modern mobile devices have onboard multiple air interfaces and increasingly powerful resources, all readily available to be shared wirelessly. Having intelligence, multiple computational and sensorial resources as well as communication capabilities is also an increasingly prevalent trend in a large number of other objects, from home and office appliances to handheld equipment and cars. In many cases, these systems are flexible and powerful enough to allow rather

Mobile Clouds: Exploiting Distributed Resources in Wireless, Mobile and Social Networks, First Edition.
Frank H.P. Fitzek and Marcos D. Katz.

advanced implementations of mobile clouds. The most obvious case is smart phones, which support programming and run of third–party applications. Tablets and more recently cameras are also portable devices with powerful processing and communications capabilities. Examples of mobile clouds being implemented on commercial mobile devices are found in [1–3]. Other examples of current initiatives working on mobile clouds or closely related concepts include Joiku [4], Opengarden [5], Instabridge [6], Waze [7] and Wi–Fi hotspot sharing a la FON. These applications were already discussed on Chapter 10.

The above initiatives can be considered to be just the initial steps of mobile cloud technology. It is worth noticing that these examples are implemented on off–the–shelve commercial mobile devices, and using conventional wireless and mobile networks. Furthermore, not even changes in the standards are required to realize these systems. All that is needed is just to develop the appropriate applications implementing the cooperative strategy required by the mobile cloud in question. New standards supporting efficiently rich cooperative interaction will encourage the development of more sophisticated mobile clouds applications. We are already witnessing developments into this direction, the standardization of Device–to–Device operation in LTE–A being the most evident example. Network architectures supporting mesh topologies and very high data-transfer rates in the local exchange of data also support efficient resource sharing within the mobile cloud. Let us now discuss how the development and further popularization of mobile clouds can be fostered. For that, we consider the role of the essential players related to mobile clouds, and how they impact on their development.

App developers: These can be considered one of the key players paving the way to the popularization of mobile clouds. It is fundamentally up to creative minds and innovative app developers to devise attractive apps for mobile clouds. A good implementation of a cool or practically useful idea into a appealing app needs no marketing, it will virally become widely popular, stimulating also the development of other ideas. Such a development pattern is well known in the gaming apps field, but it could be easily repeated in other app fields. Owing to the available programming interfaces (APIs) there is nearly no limitation for the developers. The only threat is that technology is not accessible or its access is limited. The big players in the app development scene prefer to keep some of the flexibility for themselves. Nevertheless, from a developer point of view there are APIs to work on both the mobile device and the cloud services in the network (see Figure 11.1). In order to find more effective solutions to build mobile clouds, APIs on the network would be highly desirable.

Mobile device manufacturers: The role of mobile devices manufactures is also very important as designing mobile devices proactively supporting sharing of onboard resources is one of the first steps towards realizing mobile clouds. This can be done by making possible the access of internal device resources through an open hardware platform. Information generated in onboard resources can be readily be available on a given device port to be further transmitted to other devices in the cloud, and equally, resources can receive information from other peers in

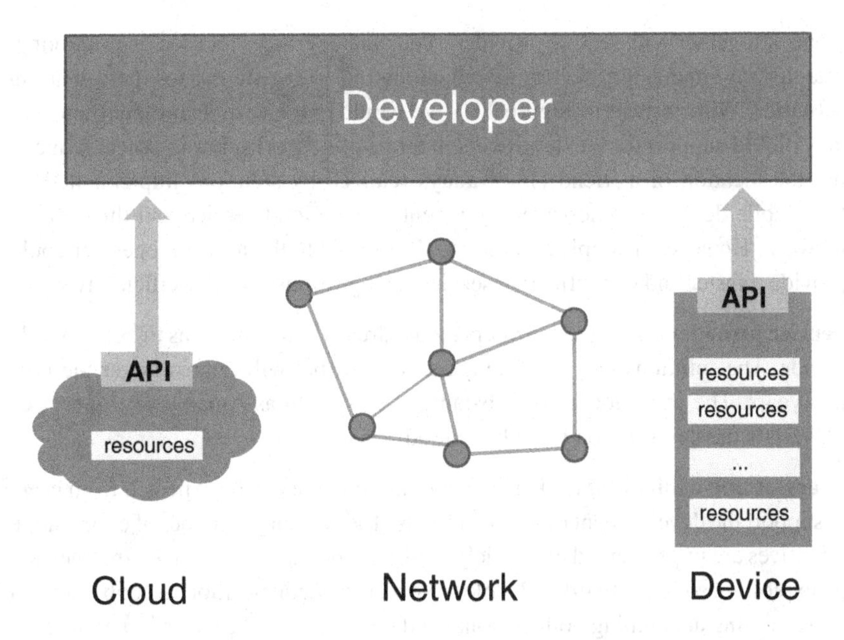

Figure 11.1 API access for developers with missing API towards networks.

the mobile cloud. Access to internal device resources, e.g., sensor readouts, is becoming increasingly available, particularly in advanced mobile devices.

Infrastructure manufacturers: Since mobile clouds can be managed not only locally (i.e., in–cloud) but also from the infrastructure side (e.g., base stations), support for this latter needs to be implemented. This includes, for instance, certain functionalities needed to assist and manage the mobile cloud operation, such as the cooperative control server discussed in Chapter 6.

Network operators: From the very initial development stages of ad hoc networking there has been a great deal of discussions on the negative view of peer–to–peer communications by network operators. Today this can be seen as a rather outdated stance, as network operators can see a number of clear advantages in having local cooperation between mobile users. Cooperative content distribution and sharing are typical examples of how network operators could considerably improve capacity and efficiency in utilization of fundamental radio resources, such as energy and spectrum. Even the simple example where a user opens his or her mobile device to cooperation (e.g., as a relaying station) shows that by doing so the user is helping the network operator to provide better service. Network operators could create incentives encouraging users to cooperate, thus benefiting users and operators themselves. Moreover, novel businesses can be created around the concept of mobile clouds. Indeed, there are many situations where network operators can be involved in the resource-sharing and -exchange process. In fact, network operators, alone or together with service providers, can create services focused on promoting, establishing, managing, securing, authenticating or registering resource–sharing activities. With respect to Figure 11.1 we would emphasize again that there is a lack of API support for the developers in terms of networks. For instance, requesting the location of a friend is nowadays realized by solutions implemented on the mobile devices, orchestrated by a centralized cloud service and through the network. However, a simple and direct API request to the network operator could provide a faster and cost-effective service, using resources more efficiently.

Service providers: Service providers today already offer solutions to build mobile clouds. The solutions they provide are sufficient, but will improve over the next few years. The provisioning of substantial APIs, such as Amazon Web Services (AWS), is the key solution to mobile cloud services.

Users: A true willingness to share is the minimum that can be expected from users to support the development of mobile clouds. Today, a large number of cooperative initiatives are implemented and widely in use, particularly through the Internet and by exploiting social networks. People have a clear predisposition to cooperate and share, aiming at attaining both personal and common benefits. This is a noticeable trend that can be capitalized on virtually by all the aforementioned players.

We can expect that a more active and widespread support towards an open exchange of distributed resources will foster the development of a wide array of mobile cloud solutions.

11.2 Mobile Clouds and Related Technology Developments

This section discusses some novel concepts as well as current developments in the field of wireless and mobile communications that are related to mobile clouds. The relationship and possible synergy between mobile clouds and each particular technology are identified, and briefly discussed.

11.2.1 Internet of Things

Internet of Things (IoT) is a vision of a hyper–connected world of objects, where virtually everything can be accessed though unique addresses. Objects are assumed to have some intelligence onboard and objects can be networked. IoT makes no assumption on the objects, they can be few or many, they can be small or big; fixed, movable or portable. Moreover, objects can interact with the environment through sensors and actuators onboard. Objects can have intelligence and processing capabilities onboard through CPU and memory and they have wireless connectivity as well. The type, complexity and performance requirements of the embedded functionalities depend on the kind of objects/things and the particular uses being considered. From our perspective, more important than the capabilities of objects is the fact that these objects can be globally integrated in a seamless manner into a communications network infrastructure, the Internet being the communications platform connecting all these objects. Having access to practically any object brings a whole new world of opportunities on how these objects can be managed, monitored, controlled or used in very many different forms. Needless to say, IoT has the potential to produce a huge impact on nearly every aspect of modern life: at home and work, in factories, transportation and logistics, for example.

The fact that IoT defines nodes with communication capabilities and the existence of the Internet as a central entity for managing communication bears a high architectural resemblance with the concept of mobile clouds, where wireless nodes interact with each other and they can also be connected to a central entity, the overlay cellular or local network. In the simplest approach IoT assumes that objects will be connected to the Internet in a centralized fashion. However, making cooperative clusters with the communication–enabled objects could be highly beneficial from the performance and energy efficiency standpoint. One could think of having a distributed architecture where things/nodes talk to each other, while a number of nodes can also be connected to the Internet, through the overlay network. Topology–wise this just follows the definition of mobile cloud, as discussed throughout the book. The main significant differences, while comparing wireless devices and conventional nodes of the IoT concept (i.e., intelligent things), are in the mobility and capabilities of the nodes (communications, processing, and battery capabilities). One possible IoT approach is the interaction of intelligent things with the nodes of mobile clouds. Thus, mobile devices, individually or as member of a mobile cloud, can opportunistically serve as gateways or connecting hubs for information generated from or going to nodes of an IoT network. Sensorial elements on the nodes/things, or for the case any resource available on board, can be also shared in the same fashion as discussed for mobile clouds. Nodes in the IoT concept are typically energy-limited, as they are powered internally by a battery, by energy-scavenging methods or through an RF–feeding loop, as in RFID approaches. Through local cooperation, as in the mobile cloud strategies discussed in Chapter 9, nodes have the potential to be connected with considerably lower energy expenditures as compared with non–cooperative cases. One IoT application example is to provide massive connection to IoT nodes, just "things". Massive information retrieval from, or broadcasting to, nodes/things consumes significant amounts of resources. Through local node cooperation, in the same manner as in mobile clouds, energy and also spectrum efficiency can be significantly enhanced.

11.2.2 Machine–to–Machine Communications

Machine–to–Machine (M2M) communications refers to technologies focused on providing connectivity to distributed nodes, symbolically machines. These machines can be considered as producing information (e.g., sensorial data, status information) as well as receiving information (e.g., being controllable). The most trivial example of M2M communications is the case where information is exchanged between nodes and a remote server. Home appliances sending energy-consumption reports to an energy provider/broker, machines being remotely operated or configured and vehicles transmitting telemetry and status information are examples of M2M technology. As with IoT, the simplest case of M2M is just provision of point–to–point connectivity, but the use of more advanced topologies paves the way to developing novel services while exploiting resources more efficiently. Nodes in M2M systems are typically assumed as not being energy–limited (i.e., driven by high–capacity batteries or connected to the power line), and with a wide range of possible processing power and sensorial capabilities onboard. Mobility–wise, nodes in M2M include from fixed ones to nodes moving at high speeds, such as cars. In general, and considering that large numbers of distributed nodes need to be connected, efficient transfer of information is required. In some scenarios where critical control and monitoring is needed, redundancy needs to be exploited to deliver reliably information to the receiving end. Node cooperation, particularly the fact that mobile clouds offer redundancy in both node– and connectivity–domains, can be readily exploited by creating alternative or parallel paths securing delivery of information. Another interesting mobile cloud approach that can be considered in M2M applications is the use of users' mobile devices as hubs for both distributing and gathering information to and from nodes, respectively. Given the mobility of these hubs, distribution of time–non–critical information can be carried out following the opportunistic principle *collect information from M2M nodes as you pass, forward information to destination server when conditions are favorable*, and vice versa in the downlink direction.

11.2.3 Device–to–Device Technology

Device–to–Device (D2D) communications is a concept being developed in LTE–Advanced, allowing direct communication between mobile devices using licensed bands of the cellular spectrum for that purpose. One of the underlying ideas behind D2D is to reduce the load of base stations by routing information locally whenever possible. As rich content traffic is becoming increasingly prevalent and social networks tend to generate spatially correlated traffic (e.g., close by users interested in the same traffic) the impact of D2D technology on offloading information from the overlay network could be significantly high. The fact that spectrum allocation for the local D2D communications is carried out in a centralized manner by the base station, and is based on licensed spectrum, guarantees that interference can be well controlled in a local environment where multiple D2D connections are established simultaneously. As D2D communications takes place over short distances, high data throughput and short processing delays can be expected. Mobile clouds in this book were in general assumed to be realized using orthogonal air interface technologies, one for the centralized access, another for the local access. However, as discussed in Chapter 4, a similar approach could be implemented with a single technology, such as the LTE–A's D2D.

D2D creates connectivity between two devices within a short–range but this concept can be readily extended to encompass multiple devices collocated in a relatively small area. Thus, in principle a network of multiple devices can be established using appropriate bands of licensed spectrum. OFDM subcarriers lend themselves well to this task, though the challenge is to carry out frequency allocations in an effective manner. This is particularly important as nodes are mobile and, as such, a dynamic system may require frequent changes in subcarrier allocations, creating a prohibitively high amount of signaling overhead. Even though the base station could effectively manage subcarrier allocations to create the required devices-to-devices links, it could be more reasonable to do it in a hybrid fashion. In fact, the base station can make a coarse allocation, e.g., a continuous subcarriers band is allocated for a given group of devices, the mobile cloud; while changes in subcarrier allocations due to dynamic changes in the cloud should be managed locally in the cloud. Several neighbor mobile clouds can also be established in this way, just taking into account that different bands are allocated to different clouds.

11.3 Promising Novel Applications of Mobile Clouds

In this section we briefly list a number of promising applications of mobile clouds. Some of them exist already today, though in rather primitive versions; others are not yet implemented.

Combining local resources: Processing power, memory, sensors and actuators, and air interfaces can be combined collaboratively in order to create more powerful functionalities otherwise impossible to obtain with individual resources. Capabilities can be augmented creating powerful virtual devices that can be used by one, few or all the users of the cloud (e.g., a virtual mobile device combining several air interfaces, with a high equivalent data throughput, CPUs computing jointly a given task such a common graphical computations needed for group gaming). Resources can also be combined to create something different than a linear increase of capabilities, e.g., creating 3D sound effects by combining loudspeakers, or directional sound capture by creating beamformers with distributed microphones. Figure 11.2 summarizes the concepts for local resource sharing, including typical sharing approaches and potential applications and services that can be developed based on these concepts.

Massive sensing: This referred also to crowd/citizen-sensing. Users contribute cooperatively with the measurements of sensors onboard of their mobile devices. It is expected that in the future there will be integrated even more sensors than today; in particular-environmental sensors (pollution, pollen, radiation, etc.) will be available. Some of them could be only activated upon the occurrence of a particular event, e.g., radiation sensors could be used just in case of a nuclear accident, such as a radioactive leakage from a power plant. Users contribute with some energy and time, by sending measurement reports regularly, or upon request from the network operators. All the measurements can be used to create two-dimensional real-time distribution plots of certain physical parameters, to be used by both authorities as well as citizens. As massive sensing needs to work without disrupting operation of the mobile communication network, it is convenient to use some of the mobile devices as hubs, gathering information from close-by devices, and relaying this information to their associated base stations. Figure 11.3 illustrates the concept of massive sensing.

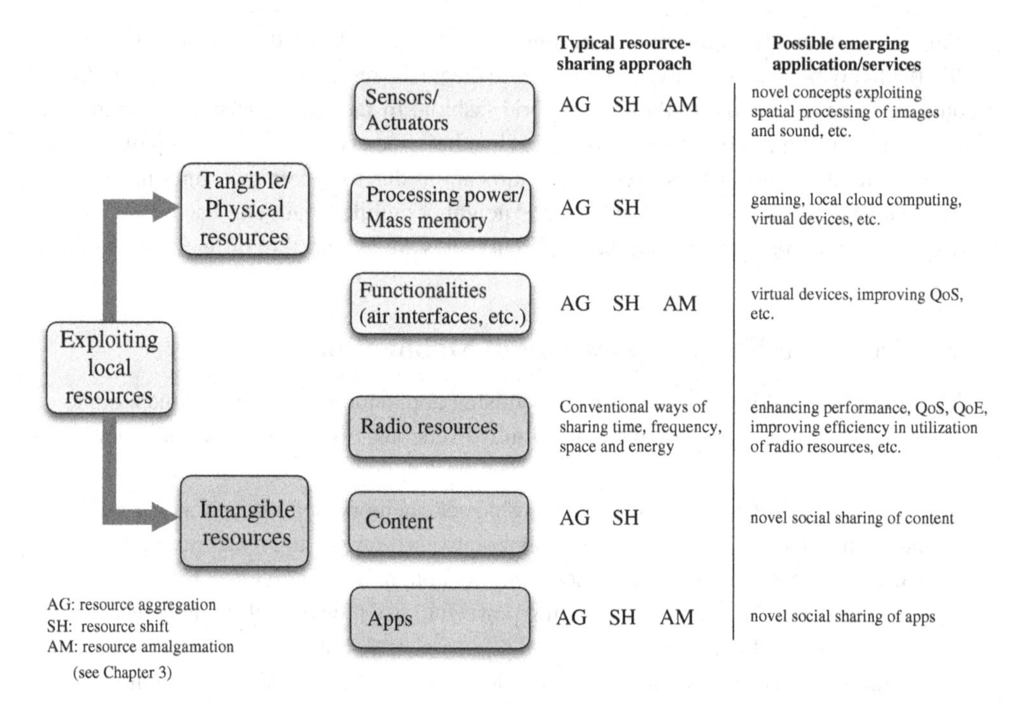

Figure 11.2 Possible applications and services based on sharing resources locally.

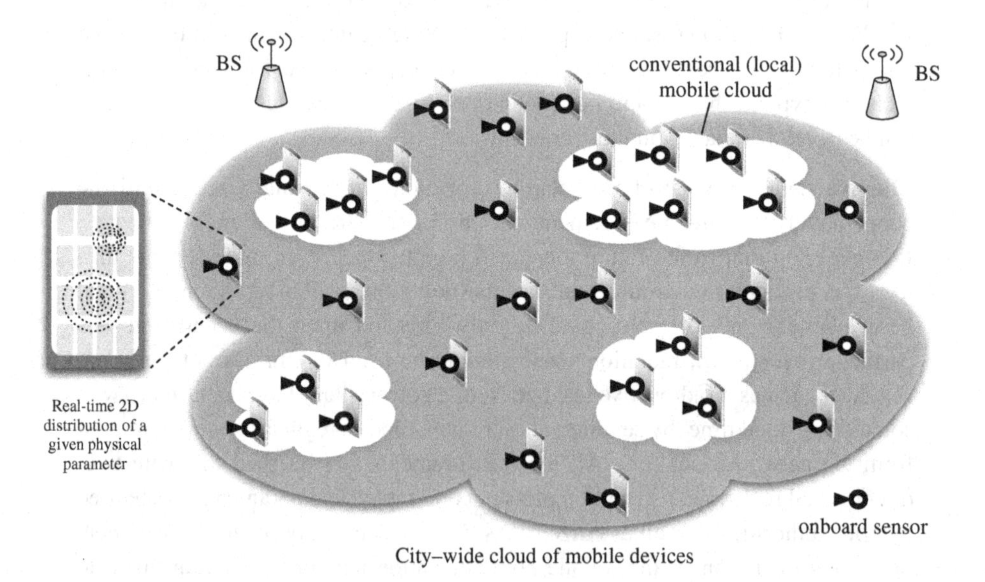

Figure 11.3 Massive sensing allows for creating real-time two-dimensional distribution maps of particular physical parameters for the benefit of authorities and citizens.

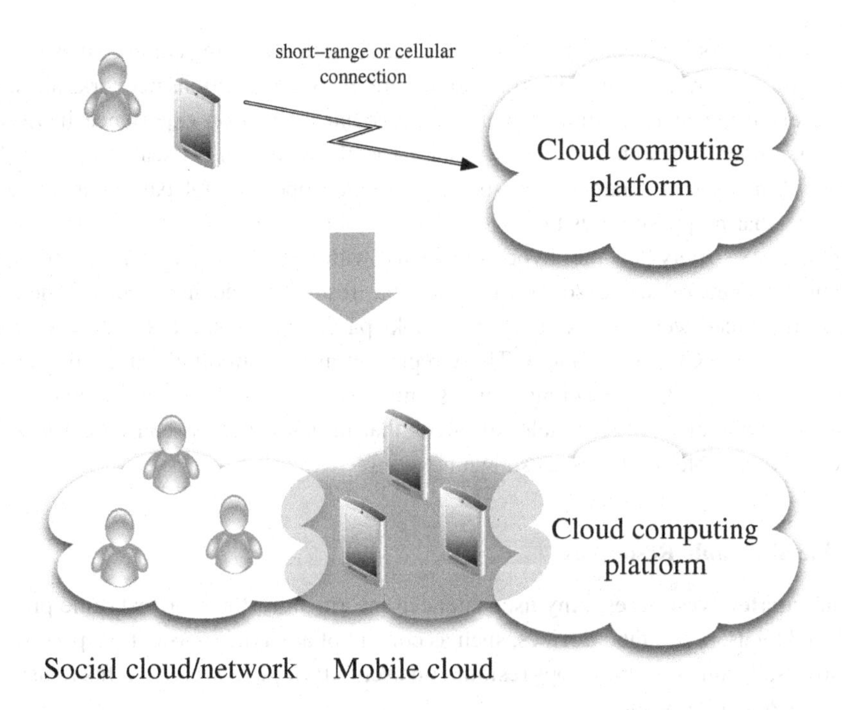

Figure 11.4 Accessing a cloud computing platform by a single user and by a social network.

Cooperative content distribution or generation: Mobile users can join forces in order to receive a similar content (e.g., video streaming) or produce it (e.g., connecting image resources to jointly produce a high-definition video out of a number of low-resolution sensors)

Cloud–to–cloud communications: A mobile cloud can be seen as the natural interface between a large-scale cloud platform (e.g., cloud computing) and a social network. The mobile cloud can provide multiple simultaneous connections between one or more users of the social network and multiple nodes of the cloud computing platform. This redundancy helps in enhancing data throughput and reliability between the social network and the cloud platform. Figure 11.4 illustrates how a cloud computing platform can be approached by a single mobile user or by multiple mobile users, part of a social network.

11.4 Resource Sharing as one of the Pillars of Social Interaction: the Birth of Shareconomy

Communications networks have already had a profound impact on how people interact. Modern communications have blurred the concept of distance by making possible nearly instantaneous social interactions regardless of people's physical distribution. Even though the Internet is the fundamental gluing factor of today's global social networks, it is ultimately through wireless and mobile networks that people can truly enjoy the sensation of being connected everywhere, wirelessly, on the move. Advanced wireless devices and communications networks allow more and more exchange of rich content, making the social interaction experience increasingly realistic. As supporting rich and almost instant social interactions is scaled up to a global level, new dimensions for large–scale social interaction open up. Of particular interest are the resource-sharing possibilities that wireless and mobile networks create on both local and global scales. As society becomes hyper–connected with fast, efficient and ubiquitous mobile communication networks, the domain of shareable resources becomes multi–dimensional. Shareable resources were discussed in this book, particularly resources available onboard mobile devices: see Chapters 2 and 3. These represent just an important but particular group of shareable resources, those residing on users' mobile equipment. One can extend these ideas to consider a wide class of shareable resources that in general terms can be classified into tangible and intangible resources, as shown next:

Intangible Shareable Resources

Information resources: Any user-intended information stored or real–time produced by users and their devices, such as content of any type (music, text, photos, movies, documents, etc.), apps residing on users' devices, as well as rights to use, access or exploit something.

Social resources: Any type of social presence that people can offer and share with other peers through communication networks, including get–together time, support, security or any form of temporal interaction creating, sustaining and reinforcing social values.

Radio resources: Common resources such as time, space, frequency and energy/power.

Knowledge resources: Any source or form of knowledge, know–how, skills, experience, education that a person can share with others.

Personal resources: Subjective conditions and emotions can be conveyed and ultimately shared through communications networks with other peers. Transfer of rich–detail information (i.e., HD live image and sound, 3D visualization) supports displaying realistically some personal feelings. With the advent of sophisticated sensors and actuators, sharing other human senses, such as smelling and touch will become a reality, making the sharing experience even more genuine.

Tangible Shareable Resources

Real–life resources: Real physical resources, virtually anything owned by anyone (e.g., objects, books, food, clothes, furniture, cars, apartments, etc.). Here we refer to *things* that can be shared, and as such they could be in principle of any type, as long as the user has the legal right to decide on their use.

Physical device resources: Any possible resource onboard a mobile device, as discussed in Chapter 2, and ultimately this can be extended to consider any shareable resource embedded on any piece of equipment. Resources include information sources (sensors) such as image sensors, environmental sensors, keyboards, microphones, position and orientation sensors, etc., information sinks (actuators) such as screens, loudspeakers, electrically–controlled mechanisms, lighting sources, etc., air–interfaces (e.g., for cellular and short–range communications, radio– and optical–based), processing power (e.g., CPU, DSP), mass memory (active semiconductor memory, hard–disks, etc.) and batteries, among others. Clearly, these resources cannot be physically moved across a cloud, and thus, resource sharing here means the capability of combining collectively signals from and to these resources.

In general most of these resources can be exchanged, moved, combined and augmented for the benefit of both a given user, a group of users, or an entire social network. How these resources can be exploited depends on many factors, the type of resources, the goal of the cooperation, the operating environment and the relationship between users, among others. Shareable resources means tradable resources in the most general case, that is to say, resources have a value that goes beyond ownership. We highlight here the social value of resources, and how useful these assets are for another peer, for a group, or for a large social network. It is clear that the value of resources can be measured with a broad array of possible figures of merit, including financial.

As mentioned before, extending the ideas of resource sharing to a very large scale creates a whole new world of possibilities. One can imagine billions of connected people, everywhere and on the move, with powerful tools in their hands, mobile devices allowing them nearly immediate rich interactions with others, globally. Widely connected people means that the resources associated with them are also connected, creating then a gigantic resource–trading platform. This trend is referred to as *shareconomy*. Figure 11.5 illustrates the concept of shareconomy as a global resource-sharing platform supported by communications networks and serving social networks. In the long run, even the meaning of ownership could be challenged by shareconomy, as it may become more attractive to share certain things, not to own them. Even though business creation can be seen as the main driving force behind shareconomy, other than monetary values certainly support the concept. People share resources for other reasons, pure altruism, environmental matters, the joy of sharing, to gain trust or improve reputation, and others.

Trading, sharing and exchanging resources has always been the base of our economy. Communications networks, and in particular mobile networks, bring almost unlimited opportunities at our fingertips, that is, resources can be shared and traded immediately, in many geographical scales, locally, or across the world. The capability of exploiting a broad range of temporal and spatial scales domains is essential when sharing and trading resources, as opportunities and businesses could arise at any time, at any place. Mobile and wireless networks will help to identify shareable distributed resources, and enable their exploitation in its multiple facets:

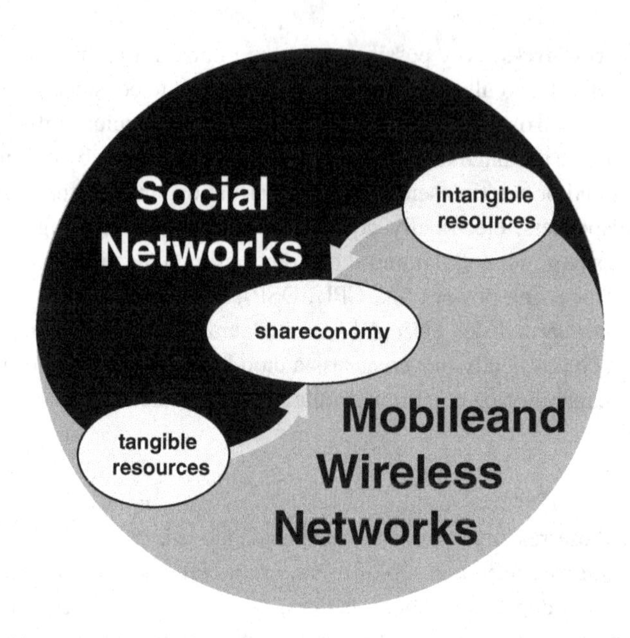

Figure 11.5 Shareconomy: The future of resource sharing in communication–enabled social networks.

sharing, exchanging, moving, trading, combining, etc. Networks will help in bringing trust and security to the process of resource sharing.

Managing efficiently these distributed resources is not a trivial task. If the resources are close by, hybrid centralized-distributed architectures such as the one discussed for mobile clouds makes sense. Thus, resources can be locally connected, but still the connections and traffic flow can be administrated via a central entity, such as a base station. If the resources are distributed over large areas, they need to be connected over their respective centralized access, and the IP (core) network, resulting in delays that could be excessive for certain applications.

References

[1] P. Vingelmann, F.H.P. Fitzek, M.V. Pedersen, J. Heide and H. Charaf. Synchronized Multimedia Streaming on the iPhone Platform with Network Coding. *IEEE Communications Magazine – Consumer Communications and Networking Series*, June 2011.

[2] P. Vingelmann, F.H.P. Fitzek, M.V. Pedersen, J. Heide and H. Charaf. Synchronized Multimedia Streaming on the iPhone Platform with Network Coding. In *IEEE Consumer Communications and Networking Conference – Multimedia & Entertainment Networking and Services Track (CCNC)*, Las Vegas, NV, USA, January 2011.

[3] P. Vingelmann, M.V. Pedersen, F.H.P. Fitzek and J. Heide. Data Dissemination in the Wild: A Testbed for High-mobility MANETs. In *IEEE ICC 2012 – Ad-hoc and Sensor Networking Symposium*, June 2012.

[4] Joiku. Joiku web page. http://www.joiku.com, 2013.

[5] Open Garden. Open Garden web page. http://opengarden.com, 2013.

[6] Instabridge. Instabridge web page. http://www.instabridge.com, 2013.

[7] Waze. Outsmarting traffic, together. http://www.waze.com/.

Index

Mobile Clouds: Exploiting Distributed Resources in Wireless, Mobile and Social Networks, First Edition.
Frank H.P. Fitzek and Marcos D. Katz.

Printed in the United States
By Bookmasters